The Day the World
Discovered the Sun

THE DAY THE WORLD DISCOVERED THE SUN

An Extraordinary Story of Scientific Adventure and the Race to Track the Transit of Venus

Mark Anderson

DA CAPO PRESS
A Member of the Perseus Books Group

Designed by Timm Bryson
Set in 11.5 point Adobe Jenson Pro by The Perseus Books Group

Cataloging-in-Publication data for this book is available from the Library of Congress.

First Da Capo Press edition 2012
ISBN 978-0-306-82038-0 (hardcover)
ISBN 978-0-306-82106-6 (e-book)

Published by Da Capo Press
A Member of the Perseus Books Group
www.dacapopress.com

Da Capo Press books are available at special discounts for bulk purchases in the U.S. by corporations, institutions, and other organizations. For more information, please contact the Special Markets Department at the Perseus Books Group, 2300 Chestnut Street, Suite 200, Philadelphia, PA 19103, or call (800) 810-4145, ext. 5000, or e-mail special.markets@perseusbooks.com.

10 9 8 7 6 5 4 3 2 1

For Penny

CONTENTS

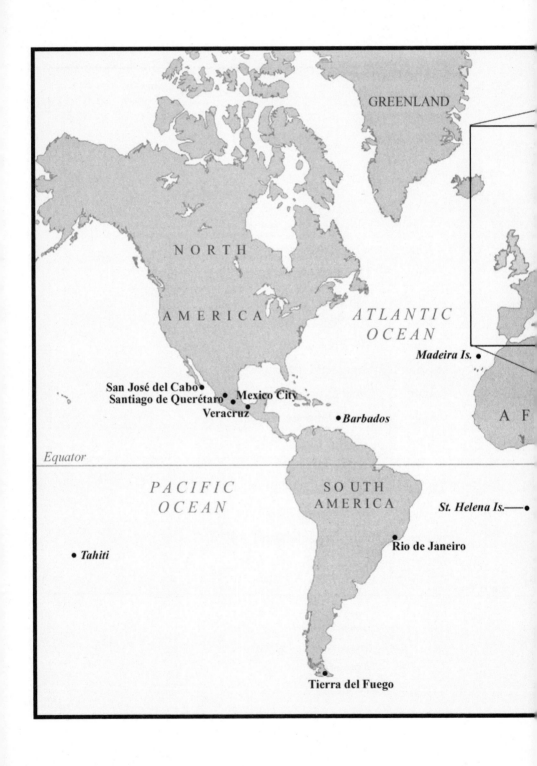

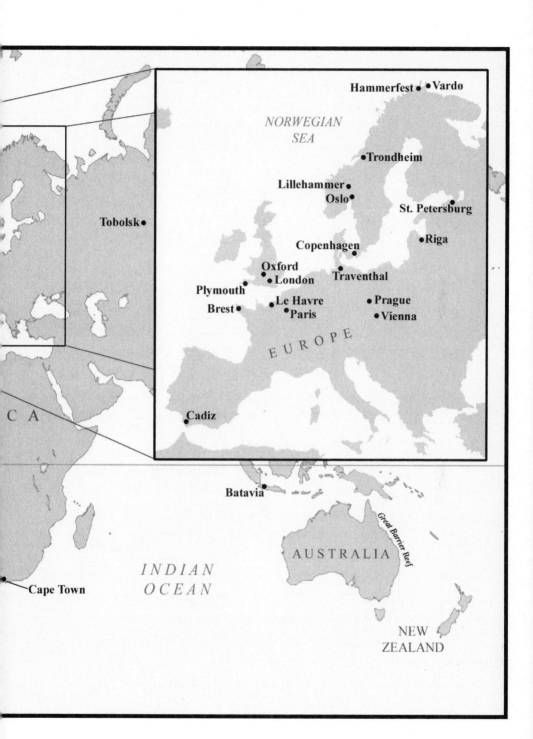

PROLOGUE

I closed my lids, and kept them close,
And the balls like pulses beat;
For the sky and the sea, and the sea and the sky
Lay like a load on my weary eye,
And the dead were at my feet.

—SAMUEL TAYLOR COLERIDGE
RIME OF THE ANCIENT MARINER

SAN JOSÉ DEL CABO, BAJA PENINSULA
May 20, 1769

Oranges, bananas, pomegranates: how could such sweet fruit go so
sour? Steady winds gusting off the Sea of Cortez did a fair job of keeping the flies from hovering over the dwindling piles of rotting food
scattered around the huts and makeshift homes near the beach. But
the flies had better places to lay their eggs. With every new day came
a new crop of corpses.

Spanish frigate captain Salvador de Medina had spent nearly seven
months escorting twenty-eight men across the Atlantic Ocean and the
whole of Mexico to arrive at a former Jesuit sanctuary near the southern
tip of the Baja peninsula. Medina had technically discharged his duty.
But his orders included nothing about a deadly epidemic, a brutal and

unforgiving fever sweeping through the local population and filling graves by the hour. Medina knew the epidemic posed too great a threat to risk staying.

Yet, the man at the center of the expedition, the French astronomer Jean-Baptiste Chappe d'Auteroche, held his ground. Chappe, as he was known, had already begun to size up an abandoned corn barn at the former Jesuit mission inland. The barn would make a fine observatory, Chappe surmised. As if consumed by a fever of its own peculiar nature, Chappe refused to hear anything more about a port eighteen miles to the southwest. Word on the ground may have been that Cabo San Lucas was free of the contagion. But Chappe told Medina that he would not risk the expedition's founding purpose simply because of some rumor.

The group's late arrival on the white sandy shores of San José del Cabo the night before had left precious little time to spare before the afternoon sky would host a sensational spectacle.

The universe, Chappe liked to explain to anyone who would listen, would soon be opening itself up for a rare inspection. Although anyone without a telescope would never notice it, for five and a half hours on June 3, 1769, a little dot would appear to cross the disk of the sun. That little dot was the planet Venus. Its shadow crawled across the sun's face, at most, only twice per century. Timing the planet's entire transit down to the second and comparing other observations of the same event from elsewhere on the globe, Chappe said, would by year's end enable humankind to discover something that had evaded it since the dawn of time—the exact physical dimensions of the sun and its planets and the distances that separated them. Venus's transit opened a brief window into the very architecture of God's creation.

This, Chappe explained, is why no mere disease could be allowed to interrupt his careful observations of the practically theological phenomenon that a fortnight later would be taking place overhead.

Having spent his first night ashore sleeping on the beach, Chappe mustered what remained of the able-bodied natives. Two leagues inland stood the mission that would serve as Chappe's observatory. Much work remained to be done, and with very little time to spare. The first job entailed hauling Chappe's delicate telescope and other scientific equipment inland—precision instruments whose every fragile inch had endured muttered curses of Spanish soldiers porting them nearly halfway across the world from ship to jungle and back to ship again.

The observatory's one widely recognizable instrument looked like the guts of a clock quartered and served up like a piece of pie. A kind of maritime priesthood wielded the quadrant with incantations and scriptures that were as mysterious as Holy Writ to most of the sailors.

A ship's navigator typically used this machine and a table of nautical charts to measure the moon and sun and sometimes stars. Through a mathematical ritual that occupied the navigator for hours on end, these measurements would then produce a crucial number that everyone at sea could appreciate: longitude.

Longitude was the most costly puzzle of its time. And astronomy was poised to solve it. Esteemed astronomers like Chappe commanded authority with royal audiences and military commanders. Commanders like Medina.

The groans of wretched men and glimpses of the cadavers they became underscored how dire the situation had become. Still, as the morning's sweat turned clammy from cooler breezes that nudged the mule train inland, Medina and Chappe remained tense allies bound by death's encroaching shadow.

Theirs was a world inching closer to discovering great secrets behind the sky. But the sun shone down relentlessly, and it forgave no one unprepared. The sun would have its day.

Chapter 1

A STAR IN THE SUN

VIENNA, AUSTRIA

December 31, 1760

Reddened hands fastened the barge to its mooring. For the past month the Danube's breezes had chilled its travelers. No more. A brisk walk from the canal bank, and Vienna—with its famously narrow streets, tall buildings, and fragrant coffeehouses—welcomed its visitors in from the cold.

Just eight years before he would travel to San José del Cabo, French astronomer Jean-Baptiste Chappe d'Auteroche and his party made a wintry landing in the capital city of the Habsburg monarchy and, with it, much of the Holy Roman Empire. His was a journey altogether of its time. The turbulent 1760s—when the Enlightenment was in full bloom but before bloody revolutions had brought the age's heady ideals down to earth—would effectively frame the world's most concerted effort to find the sun.

In transit from Paris, Chappe and his servants—as well as "M. Durieul," a Polish military man traveling to Warsaw—offloaded their Danube barge and walked through the city gates of Vienna. The crunching snow underfoot and clouds of condensed breath had become familiar companions as Chappe's party daily pressed eastward. Still, the Viennese chill could not compare to the core-consuming freeze Chappe and his servants were about to undergo. Their ultimate destination was Tobolsk, a remote town in Siberia.

In April, Chappe's colleague, the seventy-two-year-old astronomer Joseph-Nicolas Delisle, had presented a paper to the Académie Royale des Sciences in Paris arguing that Tobolsk was one of the best locations on earth to observe the coming transit of Venus on June 6, 1761. The *Mappemonde* that Delisle presented the French academy served as a sort of global menu of the most coveted destinations that teams of explorers and scientists across Europe would be risking their lives to venture to.

June 6 was the first time in living memory that the skies provided such a rare opportunity to plumb the solar system's size. Venus had last transited the sun 122 years previously in 1639, more than a generation before mathematicians had figured out the trick that enabled the sun's distance to be triangulated. Venus would provide one more chance on June 3, 1769. After that, another 105 years would elapse before Venus again passed in front of the sun.

The 1761 transit, as these scientists (known then as natural philosophers—*philosophes*) told their country's paymasters, presented the best opportunity in more than a century to get a precise fix on the sun's distance. And thanks to the planetary laws discovered by Johannes Kepler in the seventeenth century, knowing the distance to the sun allowed scientists to locate the orbital path of every planet. One measurement unrolled the blueprint to "the heavens and the earth"—what the biblical book of Genesis said God created at the universe's very beginning. It was arguably as close to knowing the mind of the Creator as anyone had yet conceived. "If we make the best use of [the Venus transits]," the instrument maker and popular science author Benjamin Martin wrote in 1761, "there is no doubt but that astronomy will, in ten years time, attain to its ultimate perfection."[1]

For a seafaring nation, discovering the distance to the sun meant advancing the frontiers of knowledge intimately connected to national security. As officials from the rival British Royal Society reminded their nation's Admiralty in a 1760 letter, Venus transit voyages required top priority attention because they constituted "the promotion of a science so intimately connected with the art of navigation as well as for the honour of the nation."[2]

For reasons that were scientific and geopolitical—if not also theological—Venus transit expeditions had become paramount. Even if they meant traveling to a remote and frigid location like Siberia. Although Russian scientists were already preparing their own expeditions to observe the Venus transit, the French Academy of Sciences had secured Chappe an invitation to make his own competing measurements of the celestial event at Tobolsk. Entrée to the Russian empire, with the empress's blessing no less, spurred Chappe and his party into the Siberian beyond.

For the next eight nights, however, Chappe would enjoy a warm bed in the comfort of one of the great cosmopolitan centers of Europe. His timing was propitious. New Year's Day in imperial Vienna was like a red-carpeted runway, providing the excuse every monied house in the city needed to strut like a peacock in full fan.

On New Year's morning, the entire society—peasants, landed gentry, middle-class burghers, the indigent—all gathered outside the imperial Hofburg palace, where the Royal Bodyguards, ministers of state, and the city's leading aristocratic families paraded through the square in flamboyant dress uniforms and courtly regalia. National identities in this crossroads city jockeyed for placement, with German soldiers marching first in line, followed by Polish soldiers and Hungarian troops in silvery uniforms and holstered sabers that glinted in the late morning sun.[3]

As a visiting French dignitary, Chappe would have stayed with the French ambassador to Vienna, the Duke de Praslin—or at least enjoyed lodgings arranged by the ambassador.

When Chappe arrived, de Praslin was caught up in a particularly busy time for a diplomat. On New Year's afternoon, families of wealth, power, and prestige gathered throughout Vienna for lavish parties that carried on well into the night. It was prime time, in other words, for a minister of state to ply and expand his network of connections.

As the afternoon shadows lengthened across the snowy streets, opulent carriages approached the city's stately homes and paused as dandified gentlemen and bejeweled ladies alighted. For a well-educated Frenchman staying in a posh part of town, Chappe didn't need a translator to understand the partygoers' chatter. Public conversations in upper-class Vienna were in French, still the language of the refined and courtly set.

Inviting aromas emanating from the kitchens of the well-heeled provided an olfactory tour of Europe: chocolate from Milan, pheasant from Bohemia, fresh oysters from Istria. Enticing music also warmed the air, as Viennese nobles prided themselves in their musical sophistication—hiring some of the finest concert maestros in the world to provide entertainment. At the time, for instance, a twenty-eight-year-old composer named Franz Joseph Haydn (four years younger than Chappe) was practically reinventing the symphony as musical director for Vienna's wealthy Morzin family.[4]

Once the New Year revels had ended, though, life at court returned to its normal state of angst. Austria was caught up in a brutal war with Prussia—putting her conscripted soldiers through battlefield abattoirs like the battle of Torgau, in which 7,000 Austrians gave their lives in one day. France, Austria's reluctant ally, wanted out of the coffer-draining conflagration, and Chappe's host was already working on his new job for 1761: convince Her Majesty to consider a peace treaty with her hated rival, Prussia's Frederick the Great.

Chappe, on the other hand, carried no such worldly baggage when he paid an invited visit to the empress and her husband, Franz I—that rare emperor who preferred to leave politics and governance to his wife.

Chappe climbed the Hofburg palace stairs to the library, where the royal couple waited to receive their learned guest.

Although Maria Theresa herself had no interest in science, the emperor did. A statesman with his own intellectual passions, Franz I showed his French visitor the biggest and most comprehensive collection of rare minerals, fossils, corals, and shells in all of Europe. The emperor's natural history cabinet—boasting 30,000 specimens collected from across the globe—even included "thunderstones" from Croatia and Bohemia. Today called meteorites, these melted miniature chunks of asteroid were at the time thought to be small pieces of earth superheated by lightning strikes.[5]

Chappe spent his week in Vienna mingling with the great scientific researchers working there. Gerard van Swieten, personal physician to the empress, for instance, shared the latest Viennese discoveries on the use of "electricity with great success in the [treatment of] rheumatism and other disorders of the like nature," as Chappe later recorded.[6]

The prince of Liechtenstein entertained Chappe at Vienna's imperial arsenal. The sixty-four-year-old Austrian military director general—who had overhauled the entire artillery, a redesign that Napoleon's generals later copied—welcomed his French guest at a suite in the military compound. Filled with state-of-the-art cannon but missing the pungent battlefield smells of death and burnt gunpowder, the prince's receiving room also had the air of a mini-mausoleum. Marble statues of Maria Theresa, Franz I, and Liechtenstein himself greeted the French visitor. Liechtenstein's school of artillery science had become one of the best in the world, so Chappe perhaps made polite conversation with his host about matters relevant to the school—explosive propulsion or trajectory calculation, for instance. Chappe also recorded accepting the prince's gift of regional fossils and local geological samples for study at his leisure.[7]

Chappe faced the bracing January winds on a visit to the rooftop observatory at the University of Vienna. The observatory's director was

Hungarian Jesuit Maximilian Höll, who had Latinized his name as "Hell." Father Hell showed his guest the telescope that would point at the sun on June 6—the Holy Roman Empire's chief witness of Venus crossing the solar disk.

Hell, a lean and intent man with a piercing gaze, discussed sky and earth with his French visitor, whose pudgier frame and baby-faced visage concealed an overriding pride at least Hell's equal. Both men of God and men of science, Hell and Chappe had as much in common as anyone Chappe would meet in Vienna. And they shared the animating passion of knowing the Creator better by better studying his creation.

Bookshelves around the observatory showcased Hell's greatest accomplishment to date. Under his direction five years before, the observatory had begun turning out its own celestial almanacs—forecasting daily sunrise and sunset times as well as regular positions of the moon, the planets, and the moons of Jupiter. These tools enabled precision navigation anywhere on the planet and inspired great pride, even in an empire that lacked a great navy.

Chappe, whose plain dress may have lost the battle of sartorial rank, retained the upper hand throughout his meeting with Hell. As a lifelong observer of the skies, Hell knew he would find no greater career-advancing opportunity than the two upcoming Venus transits. But all Hell could do this time was go to the roof and log another day. His Viennese data would be all but irrelevant beyond the walls of his own observatory. By contrast, learned men across Europe would be eagerly awaiting posts from Chappe.

The transit's extremes—the places on earth where Venus takes the longest and shortest length of time to cross the disk of the sun—produced some of the most valuable data for calculating the sun's distance. According to his colleague Delisle's all-important *Mappemonde*, Chappe was headed to one of the key stations on earth to observe the 1761 transit: the "Halleyan pole" where Venus would be taking the shortest time to cross the sun. "As the transit of Venus over the sun would not be performed in less time in this capital of Siberia than in any other part of the globe," Chappe later recalled, "it could not have been viewed to so much advantage anywhere else."[8] Chappe's observation promised to be one of the most important early measurements of the physical size of the solar system. The prospects of such groundbreaking science tantalized Hell: a single scientific adventure that could secure one's own historical legacy.

JUST OUTSIDE RIGA, RUSSIA (PRESENT-DAY LATVIA)
February 7, 1761

The screech of metal sliding over rock announced that the sleigh was stuck. No one was going anywhere. Cacophonous chatter among the Russian sleigh drivers broke the final seal on this otherwise quiet early evening. Stepping onto sled runners that had nothing to glide on, Chappe saw little else but dark. With no moon in the sky, two sources of dim light cast pale shadows on the patchy snow. To the west, the brilliant Milky Way formed a horizon-to-horizon hoop framing Chappe's entire world. And off to the southwest beamed Venus, that tiny beacon. It was setting.

For these travelers, though, the night was just beginning. Chappe's Russian translator, hired sixteen days before in Warsaw, was drunk beyond saving. "We could neither make him listen to reason nor hold his tongue," Chappe later wrote of his Siberian journey.[9] So the scientist was left to deal with the belligerent drivers himself. They quarreled and fumed, although over what Chappe had no idea. The frigid night lent a sense of urgency to an otherwise rather comical situation. Suffering frostbite—or worse—was a real possibility if they did not get help. Chappe fortunately did have a multilingual communicator with him: a single Russian ruble (equivalent to about $75 today). Showing it to his drivers, Chappe conveyed through gestures and whatever words the group had in common that he'd pay one of the drivers to go back to Riga and get help. Everyone volunteered. Each volunteer took a horse and shot off toward town, leaving Chappe and his babbling translator behind with the abandoned sleighs.

By midnight, torches and townsfolk—plus a false start involving more bribes and torn rope—had finally set the party moving again, back on carriage wheels.

As his carriage bumped along, Chappe caught a little sleep through a turbulent night. The snow grew heavier, though, and before long they wished they were back on sledges again. A morning snow squall left the carriages barely moving forward as the horses stopped every minute. To make matters worse, the baggage carriage overturned into a ditch with a tremendous crash. Earlier in the journey, when traveling by coach from Paris to Strasbourg, his group's baggage cart suffered a similar calamity. In his wreck outside Strasbourg, Chappe had jumped out of his carriage to check on the delicate scientific instruments in the cart. At the moment, though, the road-weary traveler had no impulse to dive into the snowy wreckage. Over an unpleasant din of whinnies, blows, and snorts, horses were settled down and reharnessed, and the carriage was righted. The road widened as the sun sank.

Just outside of Wolmar (today Valmiera, Latvia), the wind, in concert with a hedgerow of trees, had swept out a long line of snow banks. The coachmen carefully drove the horses through the gauntlet. The road beneath remained rocky as before, although the banks' smooth-blown surfaces fleetingly suggested fewer potholes in the road than there actually were. Then everything sank. As suddenly as a musket shot, the startled horses and lead carriage fell into a snowed-over sinkhole. Once Chappe and his companions recovered from the shock, they looked around to see the entire stagecoach had been buried. Only an opening in the vehicle's roof allowed the battered passengers to exit. The horses struggled to keep their heads above the snow, their eyes wide with panic.

The driver of the baggage carriage—which remained outside the sinkhole—jumped down and unhitched his team. The once quiet Russian roadside now echoed with a brace of shouts, heaves, and misunderstood commands. All hands now worked to rescue the animals and the buried coach from their icy interment using the only horsepower they had available. No bribes (at least none Chappe considered worth recording) passed hands this time as one of the drivers rode to town to get shovels. The group spent the better part of the day excavating the horses, vehicle, and equipment from the snowdrift.

Their cold, wet clothes, icy gear, and shell-shocked steeds made the drivers swear off carriage wheels for good. At the next town, the drivers installed runners on the convertible carriage. And at the next major posting station—the university town of Derpt (present-day Tartu, Estonia)—the travelers traded the converted vehicles for horse-drawn sleighs.

This time the weather cooperated, and no snowless patches conspired to hinder the group's progress. Sobering windchills kept normally exposed cheeks and necklines hidden beneath scarves and collars. But other than the trouble of additional layers, the party's final fortnight toward St. Petersburg was smooth as the ride itself.

As the travelers approached the colossal Russian palace that was their destination on February 13, familiar sounds that Chappe and his servants hadn't heard since Vienna pricked up their ears—conversations in courtly French. Russian empress Elizabeth, although largely uneducated, took pride in importing erudite western European culture into her realm. To her, this meant all things French: language, music, dance, art, and cuisine.[10] Elizabeth's Winter Palace—stunning and magnificent like Versailles—offered up French gastronomical delights for the starving travelers. And its halls resounded with French courtly music like the harpsichord variations of Jacques Duphly or clavichord compositions of Johann Schobert.

The opulent Winter Palace (today part of the Hermitage Museum) concealed the busy activity of its hundreds of residents and attendants— whose attention was now trained on the distinguished visitors from the west. Yet for all its comforts, the Winter Palace also harbored an un-comfortable surprise. Despite her admiration for French erudition, Eliz-abeth had also signed off on two competing Russian Venus transit expeditions to two sites near Lake Baikal—some 1,500 miles farther east than Chappe's destination. Not all of Elizabeth's court shared their empress's Francophilia, and indeed perhaps the most revered Russian astronomer of the day, Mikhail Lomonosov, did not want to see his na-tion cede to a foreigner the unique opportunity for the advancement of *Russian* science that the 1761 Venus transit provided.[11]

Nevertheless, having mollified her patriotic Russian scientists with their own pair of expeditions, Elizabeth commanded that Chappe jour-ney to Tobolsk with royal sanction. The lead horse on his team of sleighs would carry a special bell in its harness, signaling all Russians traveling the icy roads to clear the way for a vehicle of "royal post." Chappe re-quested both a top clockmaker and a translator to join his Siberian car-avan as well, provisions that were soon made. Finally on March 10, four sleighs glided eastward out of the Russian capital and into the greatest expanse of frozen wilderness the world scarcely knew.

WOODS OUTSIDE TROITSKOYE, RUSSIA
March 26, 1761

During the ensuing two weeks, the royal sleigh's interior had become quite familiar. Chappe and his four-man crew traveled straight through—stopping for just one night's bed rest in the city of Nizhny Novgorod. On this cloudy night, Chappe fell asleep as the fresh horses he'd secured in the town of Troitskoye earlier in the evening pulled their burdens through the powder. Waking up with a jolt, he had no idea how long he'd been out. And with his translator in one of the other sleighs, there was no point trying to discover such information from his driver either. His servant, the court-appointed sergeant, the translator, and the watchmaker had all been growing increasingly irritable. "They took some opportunity every day of showing their dissatisfaction," Chappe later recorded.[12] Recovering from the groggy haze of waking up still exhausted to darkness, Chappe discovered that the three other sleighs in his party were nowhere in sight. He was now on his own. Still unsure whether a dream was getting the better of him, Chappe felt the grip of a sudden adrenaline rush. His disgruntled comrades, he realized, may have abandoned him to the wilderness. "The horror of my situation will easily be conceived," Chappe wrote, "when I found myself alone in one of the darkest nights, at the distance of fourteen hundred leagues from my native country, in the midst of the frosts and snows of Siberia, with the images of hunger and thirst before me, to which I was likely to be exposed."

Chappe called out to his driver to stop the sleigh. Stepping out of the tiny cab, he stood amid the glowing blackness of the cloudy night's unilluminated snowscape. He shouted the names of his four fellow travelers into the void. Not even an echo replied.

Grabbing two pistols from the sleigh, Chappe set out along one of the trodden paths that enabled him to walk without snowshoes. Eyes adjusting to extreme dark, he could make out gross features like individual trees and the clear pathway ahead of him. In his pulse-pounding, heightened state, Chappe walked straight into the woods. He knew better, but he did it anyway. Chappe stepped off the compacted trail and— one free-falling instant later—into the shivering embrace of a snowbank. Now up to his shoulders in the frozen stuff, he exhausted himself struggling back onto the path. Catching his breath and then digging his pistols out of the snow, Chappe raced back to the sleigh to warm up. He spent most of the night methodically exploring his immediate surroundings till the windchills got the better of him and sent him back to the fur blankets in his sleigh's enclosed cab.

Within a few hours, he'd discovered a source of light in the distance. It was a farmhouse. Making a reconnaissance mission to the house, Chappe peeked in the door to discover four familiar people sleeping on the shack's floor. Next to them lay some girls, also asleep. "They seemed all to be in great want of rest," he recalled. Creeping across the farm-house floor to his servant, Chappe awoke the one Frenchman in his party without rousing the others. The boss was of course angry at his employees' reckless insubordination. But he left the house with his ser-vant "as quick as I could, for I was unwilling they should discover how rejoiced I was at finding them again."[13]

The next morning, the reunited crew set out on the road again—all sleds and all occupants present and accounted for. In broad daylight, the roads and paths revealed themselves to be slender traffic lanes that were becoming too narrow for the travelers' tastes. Two-way traffic often couldn't fit on the road side by side. Ideally, this fact shouldn't have mat-tered. With its lead horse sounding the royal post bell, Chappe's sleigh commanded right-of-way, with oncoming traffic moving to one side. But they weren't anywhere near St. Petersburg, and in the wilderness the royal post law was just another irrelevant nicety of court.

As the roads to Siberia narrowed, Chappe began to tense up when-ever he'd see a sledge approach. Some drivers would give the eastward bound party as much room as the road allowed. Others not as much. One incident crossed the line. The oncoming driver barely bothered to move his sledge out of the way, and Chappe's driver was evidently grow-ing tired of Russian scofflaws. For a teeth-clenching moment, the two vehicles looked as if they might collide. But the horses on each side clearly had no interest in ramming each other head-on. A physics ex-periment was avoided, however narrowly. But before the offending driver could get close enough to meet with Chappe's scowling expres-sion, a protruding arm of the other sledge's shaft rammed into Chappe's cab. Chappe's cab lost the joust. The sleigh's covering, Chappe wrote, "was carried away with so much force that I should certainly have been killed if the stroke had lighted upon me. This last shock completed the destruction of my [sleigh]. I now remained without any covering, ex-posed to the severity of the cold air."[14]

The three other sleds in the team also needed repairs, although none as direly as Chappe's. Consulting a regional map with his interpreter and sergeant, Chappe learned that a day's journey would bring them to the country seat of Pavel Grigoryevich Demidov, a scion of one of the richest families in Russia—and a friend of the transit expedition. A scientific amateur with a passion for botany, Demidov had given Chappe a letter in St. Petersburg commending the Frenchman to his family estates, should he need assistance during the journey into Siberia.

Demidov's Solikamsk riverside estate in the northern Ural Mountains provided comfortable refuge for a stranded traveler. Hosting a dozen greenhouses containing more than four hundred different species, Solikamsk turned out to be more than just a shack where a man could fix his sleds. "These were full of orange and lemon trees—and contained likewise all the other fruits of France and Italy, with a variety of plants and shrubs of different countries," Chappe wrote.[15] The chief mistress of the household said that the letter Chappe carried put her under orders to treat the distinguished visitor's every request as if it came from Demidov himself.

The estate's mechanic told Chappe that he'd need at least three days to fix the broken sleds, freeing the travelers to enjoy comforts unknown since Vienna. Chappe went to the greenhouse. With 10 degrees Fahrenheit outdoor breezes buffeting the window panes, the humid, temperate climate—tinged with the welcoming scent of citrus—was all the oasis Chappe could have asked for. Moreover, the greenhouse gardener was something of a philosophe himself. Demidov had, Chappe learned, encouraged the gardener's omnibus talent by creating a small science, mathematics, and philosophy library for him. The French visitor explained his Venus transit voyage to the eager audience—as comforting as any warm breeze of greenhouse air. Excited at finding a kindred spirit, Chappe gave the gardener one of the two barometers he'd made to replace the device that the weeks of travel over craggy roads had destroyed.

On the morning of March 31, Chappe wrapped himself in a fur nightgown and, with his servant in tow, took a carriage to the estate's sweat lodge. Upon opening the lodge's creaky door, Chappe walked into a cloud that he thought was smoke from a bathhouse on fire. He fumbled for the exit and made his way back into the frigid winter air. Chappe heaved a cloudy breath, doubled over, as another of the estate's staff excused himself and opened the door through which Chappe had hastily departed. Chappe conversed with his servant, who explained that the "smoke" he'd taken such hasty exception to was mostly steam. Chappe ran to his carriage, grabbing a thermometer he'd brought along for just this purpose. Always a man of science, he now reentered the lodge to investigate the environment. His servant also walked in, disrobed, and sat down. His boss's giddiness, he said, would abate if he just gave himself a few moments to relax and acclimate himself to the new surroundings.

Chappe tried. But, for starters, the stone floor and seats were uncomfortably hot. Chappe looked at his thermometer, which read 60 degrees Celsius (140° F). He got up from the hot seat too quickly, and the next thing he remembered was coming to on the sweat lodge floor surrounded by the shards of his broken thermometer. At first he didn't move. Then, from his prone position in the coolest part of the room, Chappe ordered one of the servants to throw water on him. But the dousing didn't calm the visitor's nerves. It just made him wet.

He knew he had to leave. But how to get up and go without getting up? "Attempting therefore to put on my clothes with my body bent, while I was wet, and in too great a hurry, I found them too little for me," Chappe later wrote. "And the more eager I was, the less able I was to get into them."

So he grabbed his fur nightgown, trailing bits of clothing behind him, and ran to his waiting carriage. At Chappe's command, the driver hurried back to the estate as quickly as possible. The now embarrassed guest ran to his bed. The house's headmistress was, of course, startled to see her esteemed guest in such a frenzied state. She ran up to Chappe's room and offered him some tea. He demurred.

"She gave me to understand by the Russian sergeant who began to know a little of French that I had not stayed long enough at the baths to have been sufficiently sweated," Chappe recalled. "And that it was necessary I should drink the tea to promote perspiration."[16]

Tobolsk, Siberia
April 1761

The journey's final leg bogged down as March snows melted into April slush. On April 10, the sleds crossed a final river on ice that was already underwater. The trip from Paris had consumed nineteen weeks and almost twice as many carts, carriages, sledges, and sleighs.

Approached from the west, Tobolsk looked like two cities. One sat perched on a prominent hilltop near the confluence of two rivers—the Irtysh and Tobol—that wind through the town. The other was everywhere else, in the fields and floodplains below. The entourage, driving through the western outskirts with the lead sled's post bell clanging out its imperial mission, drew locals out of their cottages. Tobolsk residents may have been accustomed to traders and trappers from the east, but actual westerners—bearing the empress's imprimatur, no less—were a rarity.

It is scarce possible to walk along the streets in this city on account of the quality of dirt there is even in the upper town. There have been foot-ways made by planks in some streets, which is the general custom in Russia. But they are kept in such bad repair at Tobolsk that you can hardly venture out except in carriages.[17]

Chappe's caravan climbed the hill to the Siberian mansions of Tobolsk's leaders. Ascending the town's central prominence provided an overview of the harsh spring thaw. The Irtysh River, which surrounds the eastern part of Tobolsk in a U shape, breached its banks in places and threatened to engulf the poorer, lower-lying regions of the city.

The crumble of icy gravel beneath the sleds' runners came to a final stop at the governor's residence. The foursome, whose wobbly legs reacquainted themselves with steady ground beneath their feet, climbed down.

The governor's eldest daughter approached Chappe and kissed his hand. Unsure how to respond, Chappe learned by the time the third daughter took his hand that he was expected to mirror her gestures and make the same kissing motions simultaneously. The governor told Chappe, through the interpreter, that he'd begun to worry that the team would not make it before the spring thaw, when travel slowed or ceased entirely.

But worries could now rest, Chappe explained. The best observers—and instruments—this region had seen had now arrived. And all that remained was to find a suitable location for an observatory. Then, with an enlisted crew of local laborers at hand, the Frenchman only needed to build.

Chappe located a hilltop three-quarters of a mile from Tobolsk and, with just twenty-six days till the transit, set out to create the site where he hoped to make history. No one suffered for daylight in this northern latitude. In early May, the sun rose at 5:30 AM and set at 9:30 PM. By transit day, June 6, daylight would trim the night by another hour on either end. Chappe unpacked and tested his equipment: the quadrant for measuring angles on the sky, his telescopes and the elaborate mechanical mounting that kept them trained on the same star or planet even as the earth's rotation made it "move" through the sky throughout the night.

Chappe supervised construction during long stretches of the day while fixing his instruments after their 3,400-mile marathon of gear-grinding and lens-scratching shocks and bumps. Chappe also had plenty of business to attend to in the build-up to the Venus transit. The astronomer had set up his clocks in the observatory building and tuned their accuracy down to the second. He further tested that his telescopes and quadrants could detect and precisely track a small shadow crossing the sun's surface.

Chappe also needed an exact fix on his observatory's latitude and longitude. To find Tobolsk's latitude (the number of degrees from the equator), Chappe measured the angle between the horizon and the sun at its highest position in the sky: noon.[18] He performed similar "altitude" measurements for the well-known stars Mizar and Caph when they crossed the same meridian line at night.

Measuring such angular distances involved using an instrument as commonplace as the compass on a ship: the quadrant. Named after the portion of a full circle that it subtends (one-fourth of a circle or 90 degrees), the quadrant was like a protractor for the sky. Measuring altitudes with a quadrant—or its cousin, the sextant, which using a clever set of mirrors enabled measuring distances up to 120 degrees—involved first pointing the viewfinder at the horizon. Then holding the right side of the instrument in place (typically using a stand), one swung the viewfinder up to find the sun or star being sited. An arm extending down from the viewfinder pointed to a semicircle hanging beneath the finder containing hash marks that read out exactly how many degrees separated the two. (Like an hour, each degree is subdivided into 60 minutes of arc, or "arc minutes," each of which is divided into a further 60 arc seconds.)

So, for instance, Chappe observed with his quadrant on May 27 that Mizar—the second star from the end of the Big Dipper—was 87 degrees, 57 arc minutes, 15 3/4 arc seconds from the horizon. His star charts told him (with some additional calculations) what Mizar's altitude would be if he were at the equator. His latitude was then the additional degrees, minutes, and seconds between Mizar at the equator and Mizar's altitude in Tobolsk.[19] The Mizar and Caph measurements both told him his Tobolsk observatory's latitude was 58 degrees, 12 arc minutes, and 22 arc seconds, give or take an arc second. According to his solar measurement, his latitude was 58 degrees, 12 arc minutes, and 13 arc seconds. Either way he was close but not quite spot-on, 39 arc seconds or about three-quarters of a mile off the actual latitude of Chappe's observatory site.[20]

Calculating longitude, though, was the real trick. French astronomers typically relied on careful timing of the motions of the planet Jupiter's moons to find longitude. Noting the exact moment (local time) when any of the four prominent Jovian moons passed behind or emerged from the planet provided an ersatz celestial clock. Nautical almanacs like the Parisian *Conaissance des Temps* contained predictions of when (Paris time) the Jovian eclipses would take place. An astronomer in the field could then, ideally, look up when (Paris time) the Jovian moon was predicted to disappear and reappear from behind Jupiter. The difference between local time and Paris time for any given Jovian eclipse, then, represented the longitude difference between Paris and the observer's location.

However, Chappe also happened into an even better longitude trick: a partial eclipse of the sun passed through the region on June 3. Chappe took down exactly when it began and ended. Ultimately he could use this number to calculate Tobolsk's longitude down to at least a few arc minutes.

None of his technical accomplishments meant a whit to anyone around him. Other than the governor and a few of his aides, practically no one in Tobolsk knew what the newcomers were doing on the hilltop near town. None of the hired hands knew what mysterious deeds lie concealed behind the intricate machinery that the strange man from afar so carefully attended to. All the locals had to judge by was the stranger's peculiar behavior.

A contemporary engraving based on a sketch by Chappe's artistic collaborator Jean-Baptiste Le Prince depicts what was probably a familiar sight at the observatory. In the image, the observatory's tall wooden front door was swung wide open with a rapturous Chappe peering through the sights of his astronomical quadrant, set squarely in the middle of the barn-like structure's opening. Through his instrument the Frenchman viewed lightning discharges during a thunderstorm. Meanwhile, local officials and assistants were huddled around the door frame's edge, as if taking shelter from the possible wrath of God that could be unleashed by such a close and inquisitive inspection of nature's wonders.[21]

Chappe did most of his instrument calibration during brief stretches at night, training his telescopes on the moon and prominent stars. His living quarters, though, were in town. Some mornings Chappe returned to his bunk in town as the 5:00 AM sunrise colored the awakening city in pink and orange. The philosophe paid little attention to clothes or grooming, not even presenting token efforts to match the ostentatious rank his royal post had given him. Already every bit the foreigner in a foreign town, Chappe recognized the suspicious glances his grizzled appearance cultivated. To everyone but the lofty folks on the hill, Chappe was persona non grata.

From his group's translator, he learned that some in town had thought Chappe was a wizard preparing to commune with the skies or some spiritual realm. Some who'd heard about the purpose of Chappe's observations whispered that the wizard was preparing for the end of the world—which, naturally, would arrive when Venus crossed the solar disk on June 6.

The storm clouds of spring gathered as the already overflowing Irtysh River hugged Tobolsk. On one fateful spring day, drenching rains and thunder rumbling through the hills gave the French visitors a second welcome. Timber crashed to the ground as the Irtysh breached its banks and drowned sections of Tobolsk in floodwater. Refugees seeking higher ground brought reports of townsfolk drowning in the flood.

Some superstitious locals didn't wait for the rains to abate to begin pointing fingers at the strange man from out of town. As if to punctuate the fury from above, mudslides broke off chunks of Chappe's mountain, sending clumps of hilltop booming into the plains below.

The rains eventually receded but the Irtysh did not. From the observatory, Chappe wrote, the flooded regions of Tobolsk looked like "a number of islands scattered on [a] watery surface and extending as far as the sight." Standing on the muddy grounds of his now completed observatory, Chappe gazed at the unfortunate town below. Tobolsk's governor had already assigned a second sergeant and three grenadiers to Chappe's entourage.

The guard now tailed Chappe wherever he went—which, by consensus, would be the hilltop and the hilltop only. Chappe started sleeping at the observatory, fearing his presence in town might incite an angry mob. For this reason, some on the hill advised Chappe not to visit the observatory without his armed guard to protect him. The locals, Chappe said, "imagined they should see no end to their misfortunes till I was gone from Tobolsk."[22]

Chapter 2

THE CHOICEST WONDERS

Offshore of Salcombe, England
February 14, 1760

Sharp winds and stinging rain greeted all hands on His Majesty's warship the *Ramillies*,[1] who had lost her course and fallen away from a military blockade of the French coastline somewhere in the English Channel. A ninety-gun ship of the line, *Ramillies* had seen action in seven naval battles over five wars. But now the seas were battling her. The storm had sprung gushing leaks in her hull.

Her navigator recognized a welcoming headland on the nearby Devon coastline. Captain William Whitrong Taylor ordered *Ramillies* hoved-to for repairs, assured by the promise of a sheltering bay. But as she drifted closer to land, the winds and surf only battered harder. Rocky coastline stretched well beyond where the bay's inlet should have been. The inlet was actually far west—one-third of a degree longitude west—from where Captain Taylor thought it was. In fact, the *Ramillies* had drifted into perhaps the most deadly length of shore in the entire channel. Before long, the captain realized this too. He shouted contravening orders to his commanders on deck. But the emergency course correction snapped the mainmast and mizzenmasts and shredded the remaining sails to ribbons.

Dropping two anchors seemed to rescue the moment, but the panicked crew failed to notice the dual anchor lines twisting around each other. Tension built until a snap unmoored the ship and sent it hurtling stern first into a cliff side. Seven hundred men died gruesome deaths that night. And the next morning, the Devon locals onshore ignored the bloated bodies bobbing in the surf to rescue hardtack for their pantries and stray shards of hull for their lumber piles.

Ramillies was just the latest sacrificial offering on the altar of longitude. In 1707, more than 1,400 sailors died when four British naval ships sank off the Isles of Scilly. The fleet's navigators had mistaken the craggy English archipelago for Ushant—an island that marks the southwestern entrance to the English Channel. To commemorate the Scilly disaster, Parliament ultimately passed the Longitude Act of 1714— legislation that established a princely prize (up to £20,000) for a practical method to reliably find longitude at sea.

In the generations since, no one had claimed it, though many had tried. The bureaucratic body set up to administer the prize, the Board of Longitude, had considered some bizarre proposals over the years. One, for instance, would have set up a network of anchored gunships throughout the Atlantic that fired clock-synchronizing exploding projectiles into the sky every midnight.

The board's most promising prospect involved comparing the moon's position in the sky to its predicted position from navigational tables printed months in advance. One of the prodigies working on these very tables was a West Country boy whose name would one day divide a nation.

OFFSHORE OF SALCOMBE, ENGLAND
January 9, 1761

Charles Mason squinted through the early morning light to make out the cove where the *Ramillies* had crashed ashore just eleven months before. Mason's ship, the HMS *Seahorse*, creaked the easy groan of a frigate cutting a line through the English Channel.

The *Seahorse* hadn't even been under sail for twenty-four hours. But disquieting recollections of a treacherous coastline to starboard begged the question: How soon till we set a course toward the open Atlantic? The coming day would in fact be the last day Mason could lay eyes on English land for more than a year.

Mason and his assistant, Jeremiah Dixon, were making hasty passage to Bencoolen, Sumatra (today Bengkulu, Indonesia) to observe the June 6, 1761, Venus transit.

Mason, 32, had made a reputation among Gloucestershire schoolmasters and tutors as a mathematical wunderkind. England's Astronomer Royal, James Bradley, hailed from the same county and through local connections learned about Mason's prowess. Bradley hired Mason in 1756 as an assistant at the Royal Observatory in Greenwich.

Dixon, 27, was the son of a wealthy coal mine owner near Newcastle-upon-Tyne who began his career as a surveyor but showed remarkable talent at both math and astronomy. Dixon's skills impressed a family friend, one of Bradley's top instrument makers, enough to recommend the surveyor when the Royal Society began casting about for someone to assist Mason on his 1761 Venus transit voyage.

The East India Company operated a factory in Bencoolen that, as one contemporary visitor put it, "produces some drugs, but chiefly pepper."[2] The factory town was going to be their new home and site of an astronomical observatory of their making. The company had promised to cover Mason and Dixon's "diet and apartments" in Bencoolen and "whatever else the service they are employed upon may require." Plus the company would cover all expenses for the observers' passage home. All their services expressed, as a company memo to the Royal Society put it, that this profit-oriented corporation was "extremely desirous of contributing every thing in their power for facilitating the making of observations upon the transit of Venus."[3]

Mason and Dixon, who had never met before the Royal Society paired them up for this voyage, were promised £200 each ($50,000 in today's money) to venture 14,000 miles around Cape Horn and through the Indian Ocean to their Pacific destination. Neither had traveled more than three hundred overland miles from home before. And home, for both of them, meant some heavy freight to be left behind. Mason was a recent widower, with two sons whose care he had to arrange for in his absence. Dixon was a tippler—having been kicked out of his Quaker congregation for excessive drinking just three months before.

As seamen manned the first night watch, *Seahorse* sailed through the icy waters of the channel toward a rare alignment of the moon and planets. Within 40 degrees of one another in the southwestern sky lay Venus, Mars, Jupiter, Saturn, and a sliver of a crescent moon. Both Mason and Dixon had earned their place on this voyage as meticulous watchers of the night sky. Mars shined near the horizon, south by southwest, separated from Venus by 10 degrees and 29 arc minutes. Jupiter practically sat atop Venus, a mere 3 degrees and 6 arc minutes distant in the sky. Both hugged the upper edges of the moon's crescent.[4]

The Astronomer Royal's instructions for observing the Venus transit once they'd landed at Bencoolen called for both timing the duration of the transit and making angular separation measurements with their quadrant. "Observe the first and second contacts of Venus with the limb of the sun," Mason's boss, James Bradley, instructed. "Then measure the distance of Venus from the limb of the sun to ascertain the nearest approach of Venus to the center of the sun's disk. Measure the diameter of Venus."[5]

More instructions followed about setting up and calibrating their pendulum clock with the varying day and night temperatures in Sumatra and the sun's slowly changing passage through the sky. More nights ahead would surely find Mason and Dixon getting to know each other's quirks and points of personal style in the telescope and quadrant measurements they'd need to keep.

8.5 LEAGUES NORTH OF USHANT ISLAND, FRANCE
January 10, 1761

The cry from high on the mainmast, at least as far as international agreements were concerned, represented no threat whatsoever: "Enemy sail bearing down hard to windward!" The *Seahorse* carried a scientific mission that both French and English courts heartily supported. Her captain, James Smith, had every military permission, domestic and foreign, to sail under a white flag—a universally accepted protocol for warships undertaking nonhostile expeditions.[6] Of course, white flags were also run up at times of surrender, and Smith captained a military vessel full of English seamen who knew from six years of fighting that the tide of war was now turning in their favor. As the leeward ship, the *Seahorse* had poor maneuverability and was thus the more vulnerable vessel as it rode toward the enemy.

The *Seahorse* was a "sixth rate" frigate, carrying a modest 24 nine-pound guns and 160 men—carpenters, sail makers, quarter gunners, and yeomen of the powder room—to make this small fortress function. Considered too weak for the battle line, sixth rates were typically, as a Royal Marines handbook of the time put it, "destined to lead the convoys of merchant ships, to protect the commerce in the colonies, to cruize in different stations, to accompany squadrons, or to be sent express with necessary intelligence and orders."[7]

As the French ship held its tack, it came into viewing range of the captain's nautical spyglass. A bigger and more intimidating warship came into focus. The thirty-four-gun frigate *L'Grand* was clearly not interested in discussing cosmographical matters with the astronomers onboard. "Monsieur," the snide nickname English sailors gave to French ships, had its big guns ready. Captain Smith wouldn't deny the enemy its due.

Mason and Dixon, men of philosophy and "mathematicks," had no place on deck once *L'Grand* came within firing range. Perhaps sequestered away in one of the officer's quarters or their own lodgings, the two said their prayers. Thumps from afar and loud cracks from the gun deck made every moment of the eleven o'clock hour a God-fearing one. Marines boarded the *L'Grand* in punctuated waves of assaults, while their French counterparts shouted "Allez à l'abordage!" and crossed a plank onto the *Seahorse* for tense minutes of close-quarters musket shots and bayonetting. Heard from behind closed doors, the din of screaming, shouting, moaning, vomiting, exploding, and crashing only heightened these men's focus on their delicate telescopes. With the June transit date fast approaching, and no time or place on their itinerary that allowed for replacements or repairs, Mason and Dixon could be severely injured or maimed without necessarily jeopardizing the mission. Not so their precious instruments.

Mason and Dixon's shipmates worked feverishly to better the volleys of French cannonballs bashing *Seahorse*'s timbers. However, for the instruments under Mason and Dixon's charge, this was exactly the problem. Unless an unlucky French cannonball actually pierced the boards protecting Mason and Dixon's cabin—in which case both instruments and instrument tenders were probably done for—enemy fire posed less of a threat to delicate casings of glass, wood, and metal than did the *Seahorse*'s own cannon. With every English broadside came a fusillade of case-rattling tremors and quakes. With every English broadside, Mason and Dixon had the work of ten hands spread out among four.

The next hour and a quarter was wide-eyed with what doctors of the day called "excitement of the nerves." Pulses pounding, with cold sweat dripping from their brows and chins, the ship's two scientists stayed their post and ensured no lens, gear, or eyepiece met with the same fate as bones and skulls of the fighting men above.

Then, just as suddenly as the engagement began, the French booms from beyond stopped. Groans and wails from the *Seahorse*'s injured continued unabated. But from behind a closed cabin door, the sounds of the able-bodied had turned from those animating a warship in battle to those driving a frigate under sail.

Captain Smith had more than a score of maimed and wounded men, many beyond saving. "Monsieur" may have broken off the battle, but this was still war. The *Seahorse* gave chase as best she could. But French hulls, known to be sleeker than those of the lumbering English fleet, gave *L'Grand* a natural advantage that "monsieur" used to his benefit.

Soon a battered English frigate turned around and sailed back north whence she'd originated.

PLYMOUTH, ENGLAND
Monday, February 2, 1761

The post coach from London via Bath didn't run on Sundays.[8] So the letter that arrived at Mason's temporary lodgings in this dockyard city had one day longer to steep in its rich juices. It was dated Saturday, January 31.

"Resolved unanimously," it began. "That the Council are extremely surprised at [Mason and Dixon's] declining to pursue their Voyage to Bencoolen and which they have solemnly undertaken; and have actually received several sums of money upon account of their expenses, and in earnest of performing their contract."

Dogs—even cattle and poultry—openly foraged for food in Plymouth's filthy streets.[9] The farmyard chorus of moos and clucks provided all the sonic accompaniment that Mason, not lacking in a good sense of humor, could have wanted as he read through the Royal Society's haughty post.

The society's governing council, including a new member from the colonies named Benjamin Franklin, clearly had not appreciated Mason's argument that withdrawal, under the circumstances, was necessary. Three weeks had now passed since the *Seahorse* had limped back to port. Some of the nation's finest shipwrights were repairing the battle damage.

On January 12, Mason had posted notice to his paymasters at the Royal Society in London about the unfortunate turn of events in the channel. "The stands for our instruments are tore very much," he reported. "But the clock, quadrant, telescopes, etc. are not damaged that I can find."[10] He asked what the society would have them do next. They soon learned the society's wish, as Mason paraphrased their new orders, to "do every thing in our power to answer the intention of our expedition."[11] Stay the course, in other words. Mason and Dixon were to wait in Plymouth and sail on the *Seahorse* once she was seaworthy again.

And so the Royal Society's representatives bided their time as each passing day made the mission to Bencoolen more likely to miss the transit altogether. The society should probably, considering the lengthy voyage, have initiated the expedition two or three months before it actually did. But now another precious month would pass before Mason and Dixon could log even their first bona fide nautical mile beyond English coastal waters.

The duo had been writing letters to both Royal Society officials and the Astronomer Royal begging to be reassigned to a closer destination, one they knew their ship could reach with time enough to construct an observatory worthy of their task. The eastern Mediterranean coastal city of Scanderoon (Iskenderun in present-day Turkey), they said, "will make a third point upon the Earth's disk of very great advantage to those of St. Helena [where a second Royal Society transit voyage had also been posted] and Greenwich."[12] Moreover, Mason informed his boss in Greenwich that his calculations suggested Scanderoon or another location in the eastern Black Sea could give them a leg up against competition from across the English Channel. One of these alternate locations, Mason wrote, "will answer beyond those of the French in Siberia."[13]

The two men undoubtedly were aware of the additional fact that, especially in a time of open war in the Atlantic, sailing through the Mediterranean rather than around Cape Horn drastically reduced the risk of shipwreck by storm or of another rendezvous with the enemy.[14]

And now, receiving the January 31 letter from the council, Mason had the final answer he'd been seeking. Not only did the council reproach Mason and Dixon for daring to suggest workable alternatives, the august body saw their action as something approaching mutiny. The Royal Society threatened legal action if the two continued with any more innovative thinking.

"That in case they shall persist in their refusal," the letter concluded, "or voluntarily frustrate the end and disappoint the Intention of their Voyage, or take any steps to thwart it, they may assure themselves of being treated by the Council with the most inflexible Resentment, and prosecuted with the utmost Severity of Law."

The next day, Mason and Dixon sent a short, apologetic post back to the council. "We shall to our best endeavours make good the trust they have pleas'd to confide in us," they said—adding, in a postscript, "We hope to sail this evening."[15]

Table Bay, Cape Town (Present-Day South Africa) *April–May 1761*

Two weeks on the open ocean for someone as prone to seasickness as Charles Mason might make a nonconfrontational man pine for the English Channel—even if it meant another gunboat fight. But more than two months at sea, steadily southward past the Azores and riding the trades west toward Brazil and back east again, provided daily affirmation of Samuel Johnson's quip, "Being in a ship is being in a jail—with the chance of being drowned."

Poorly ventilated and pungent with every foul odor human bodies can unleash, a frigate at sea berthing 160 men in its close quarters gave Mason and Dixon ample excuse to take the fresh air on deck. They certainly exercised such opportunities when working with the ship's master to fix *Seahorse*'s latitude (via measurement of stars and the sun against the horizon) and its longitude (via lunars) along the way. Hours spent measuring stars and planets and calculating the ship's position probably constituted the best moments of the day that the two mathematicians spent on an otherwise wearisome passage.

As winter turned to spring—and, descending into the Southern Hemisphere, toward cooler weather again—the prospects looked progressively worse for anything approaching a further 6,000-mile voyage to Sumatra. The Dutch East India Company port of Kaapstad (Cape Town) represented the last hope for setting up their temporary observatory in anything resembling civilization.

An English ship dropping anchor in late April in Cape Town's Table Bay[16] already told company officials plenty. The stormy *quaade mousson* (Southern Hemisphere winter) was fast approaching. Either the visitors were stocking up for a hellish passage into the Indian Ocean, or they were somewhere they hadn't planned to be.

An exchange of salutatory cannon fire between fort and ship opened the dialogue. The last time Mason and Dixon's ears had rung with such thunder, of course, an enemy threatened to take away their phenomenal career-making opportunity. This gunfire, on the other hand, represented the opening salvo in a backup plan that might give the Venus transit back to them.

Dutch officers boarded *Seahorse* to learn of the circumstances that brought her to their shores. Captain Grant, prideful mariner that he was, couldn't tell the Dutch officials that *Seahorse* had been bested by a French warship. Instead, he explained that circumstances outside their control had forced *Seahorse*'s hand.

As luck would have it, Cape Town's governor was no stranger to scientific expeditions. Ryk Tulbagh,[17] who'd overseen an era of reform during the ten years he'd reigned, had in 1751 welcomed the French astronomer Nicolas Louis de la Caille for a two-year sojourn that included compiling the world's most detailed sky map of the Southern Hemisphere. Tulbagh also counted famous natural philosophers among his personal friends, including Swedish botanist Carl Linnaeus. The legendary Swede had once written to Tulbagh that, even if he could switch places with Alexander the Great, he'd still rather be governor of the Cape. "The Beneficent Creator," Linnaeus wrote, "has enriched [Cape Town] with his choicest wonders."[18]

To his credit, Tulbagh welcomed the scientific expedition with open arms, unheralded and unannounced though it was. And so Tulbagh's star-shaped Castle of Good Hope, close by the city's main dock, became the first unnatural wonder to greet the visitors as they offloaded their gear.

A second spectacle awakened the senses upon passing the stone fortress. As one contemporary put it, the bulk of the city was located "a good musket shot to the west of the castle."[19] And walking toward Cape Town's sea of thatched roofs and the company's sumptuous garden beyond, a brutish contradiction came into view that visitors from afar had fresh eyes to see.

While admirers like Linnaeus praised Cape Town as "paradise on earth," it was also a city built, maintained, and run on an often unspeakably brutal institution of slavery. Gibbets around the town's periphery exhibited condemned slaves' maimed corpses and body parts.[20] Mutiny onboard a ship might lead to executions, but the punishment of swift death looked merciful compared to the drawn-out and horrific ends revealed by these vulture-chewed charnel remnants.

From the porters who carried the expedition's gear to the servants in Tulbagh's entourage to the manpower that kept the bulk of this company town moving, the city's slaves also served as living mementos of conquered cultures to the north and east. Tribesmen from mainland southern Africa and Madagascar, as well as Indians, Ceylonese, and Indonesians, all made up the city's shackled population. Many slaves eked out their miserable lives in an overcrowded lodge near the castle, where the inhumane conditions (in some years the mortality rate was 20 percent) would have been unacceptable even in the Dutch East India Company's zoo farther inland.[21]

The zoo comprised a small portion of a forty-three-acre rectangular garden that practically defined the city. Sailors the world round may not have visited or even known of the Compagnies Garten. But if they sailed the Cape regularly, they had certainly feasted on its output. Mutton, beef, fruit, vegetables, bread, wine, and fresh water all flowed like manna—ensuring visitors like Mason and Dixon could scarcely want for good food and drink during their stay. And just east of Cape Town's cornucopia lay the company stables, beyond which city officials allowed Mason and Dixon to set up.[22]

As the rainy season approached, the men turned toward their final pretransit duty—constructing the observatory. The first two and a half weeks of Mason's meticulous Cape Town observer's log recorded just one night of stargazing. Had the duo wanted to calibrate their instruments, the skies didn't permit. It was cloudy, Mason notes, "nearly all the time." Mornings especially, he said, "are very subject to thick fogs, which lie till 9 or 10 o'clock. But I hope our intentions will not be finally disappointed."

Mason wrote back to his paymasters at the Royal Society on May 6 that Tulbagh had "supply'd us with necessaries for building an observatory, but the Dutch are so slow and so few speak English that I was very doubtful of getting it completed in time."[23] So instead, Mason and Dixon hired the *Seahorse*'s six carpenters—men trained to patch leaky hulls and masts shot through with cannon fire—to build their tiny wooden shrine to the stars. The expedition's account books note a hefty cash outlay of £12, 6 shillings, 6 pence ($3,000 in today's money) going toward timber, a rare and expensive commodity on the Cape. The carpenters clearly knew how essential their services had become, and they charged accordingly. Their ten days of labor cost another £12, along with a hefty sum of £13, 16 shillings ($3,400) covering the costs of "victualing."[24]

By May 18, the well-fed men had completed the twelve-foot-wide structure and set the observers' expensive pendulum clock onto boards sunk four feet in the ground. The whole structure was less than nine feet tall, with three feet occupied by a conical roof. The roof rotated with a sliding aperture that enabled both telescope and quadrants to be pointed through the hole. Heedless visitors to the cramped space found ample opportunity to stub their toes and slam their head on the low door.

Cape Town
June 5, 1761

For five weeks, Mason glared at the skies. Clouds and rain cursed the explorers practically every day and every night when they tried in vain to point their telescopes at the planets and stars. The observatory was now fully ready, but the finest instruments in the world meant nothing if the weather refused to cooperate.

At dawn the next morning, June 6, the sun would rise with the Venus transit under way. Cape Town provided a window seat for only the second half of the planet's progress across the solar disk. The three hours immediately after sunrise would see Venus's ink spot shadow crawl up the rest of the sun's face and then disappear from the top edge, like a dog tick climbing and then falling from a lamp. This celestial spectacle—in conjunction with other transit measurements elsewhere on the globe—could still give the Royal Society the numbers it needed to triangulate the sun's distance.[25]

On the other hand, the society would hardly look kindly on a Venus transit mission to Bencoolen that never reached Bencoolen and then, even from its makeshift secondary observatory in Cape Town, returned home with nothing but tales from afar. If the skies on the morning of June 6 were like the cloudy skies every other morning during their Cape Town stay, these Royal Society hired hands could probably expect more love letters from the governing council like the one they'd sent to Mason on January 31.

But on the night of June 5, their fate took a turn. The clouds began to dissipate. Mason and Dixon, for only the second night of their entire Cape Town stay, were now free to do their job.

To the east, the star Antares—brightest star in the constellation Scorpio—rose above the jagged peaks of the Stellenbosch Mountains.[26] The occasional whinny from the Dutch East India Company stables broke the studied silence of the most serious night in these men's lives. Mason trained his scope on Antares over the course of the night for thirty-two separate measurements of its height above the eastern, and later the western, horizon. At Mason's each recording of the star's altitude, Dixon read off the clock down to the half second. Clouds obscured four of the thirty-two measurements, however, as if reminding the astronomers how close to precipice they remained.

With the night chill and the seasonal offshore Cape winds buffeting the round-topped shanty, the observers took turns rotating their tiny dome to the other star they monitored throughout the night: Altair. Known to sky watchers as one of the three vertices in the "summer triangle," this late-rising star—whose progress through the sky Dixon closely monitored—glimmered to the north over Table Bay.

Dixon completed his final measurement of Altair in the northwestern sky as the first fingers of dawn colored the bay's anchored ships in shades of marzipan. Mason and Dixon rotated their observatory roof toward the east. And so at 5:45 AM on this morning of sweaty palms and adrenaline, they cast their fate with a turgid atmosphere.

The sun should have risen, Venus's transit already under way, glazing the top of Tygerberg Hill with the day's first hint of sunlight. But Mason logged in his journal, "The sun ascended in a thick haze and immediately entered a dark cloud." The skies had turned sour again.

JAMESTOWN, ST. HELENA ISLAND
February 1762

When he met them after their respective Venus transit experiences, Nevil Maskelyne had had little interest in hearing Mason and Dixon's stories of bad weather.

After spending another three months in Cape Town, the two explorers had arrived in October in the coastal town of the British-occupied island of St. Helena—one of the remotest islands on the planet, 1,100 miles west of the nearest African coastline and 1,900 miles northwest of the Cape. They'd come to assist Maskelyne, a fellow Royal Society explorer.

The twenty-nine-year-old Anglican curate had been holed up on this seven-mile-wide shard of volcanic rock for more than a year. Maskelyne's assistant, the navigator Robert Waddington, had left his boss behind soon after June 6. (The Royal Society had hired Waddington only to help with transit observations.) In fact, Waddington enjoyed a more productive time returning to England than he did during his St. Helena stay. Sailing home, Maskelyne's assistant so impressed the ship's captain with his lunar navigational skills that the captain refused any payment for Waddington's passage.

On the other hand, at dawn on June 6, Maskelyne and Waddington experienced much the same nervous anticipation that Mason and Dixon had known. Clouds flirted with the sun, occasionally blurring it or hiding it altogether. But Maskelyne and Waddington had also snuck in some measurements of an immersed Venus's varying distances from the sun's edge. Yet all was trivia unless the St. Helena observers could secure one crucial number: the exact time (down to the second) when Venus's silhouette touched the inside of the sun's edge. (In 1761 only the latter half of the Venus transit was visible in the south Atlantic. So lacking any ability to measure the complete duration of the transit, a different method of computing the sun's distance became necessary: one that compared from different observing stations on the globe the exact time of a single moment in the transit—in this case the instant when Venus first began exiting the sun's disk.)

Just a few minutes before Venus's shadow began to exit the sun, the whole proceeding disappeared from view. Maskelyne and Waddington watched, dumbstruck, as a cloud killed the purpose of their entire transatlantic adventure.

Mason and Dixon, on the other hand, had experienced just the opposite. Clouds obscured parts of Venus's middle passage across the sun. But Cape Town's skies cleared when the crucial moment arrived. Mason wrote "very clear" in his record of the moment of internal contact between Venus and the sun.

Maskelyne—none too modestly referring to himself as "our astronomer"—later wrote, "The cloudiness with which the island of St. Helena is so frequently infested . . . unfortunately deprived our astronomer of the important observation of the exit of the planet from the sun's body." But, he added, "on this trying occasion he is said to have born his disappointment with so much fortitude as to have said he hoped to meet with better weather to observe the next transit."[27]

Endowed with a squarish and sometimes snarly face, Maskelyne commanded an intimidating presence only heightened by his determined and singular focus. Maskelyne's claim of meeting such crushing failure with pluck and renewed determination may not be completely outlandish. Perhaps he ultimately was excited for Mason and Dixon's success. But the air surely weighed heavy at Mason's first meeting with Maskelyne at St. Helena.

Mason and Dixon's October arrival at Jamestown—a small (British) East India Company island way station—came with the fairer weather of late (Southern Hemisphere) spring. The rough, southeasterly winds of the open Atlantic mellowed into offshore breezes as Mason and Dixon's frigate rounded the bay and tucked into Jamestown's half-moon-shaped cove. The distant din of a ghost town awakening greeted the visitors, as country farmers flocked to port to engage in the chief economic activity this shut-in community could claim: trading and haggling with sailors from visiting ships.

The dock became a mini marketplace hawking fragrant baskets of produce—yams, bananas, grapes, and figs—and squawking hens and geese in exchange for calico, silk, or sugar. Every day as Mason came down to the waterfront to measure the tides (part of his daily routine for Maskelyne), he came to learn the strange rhythms of an economy moored to the whimsy of passing ships.

He also came to know life working for a man at least as prone to drink as his underling, Dixon. The reverend's liberal purchase orders of casks of rum and claret and hogsheads of old porter[28] for his St. Helena stay reveal a more liquid-based life of the spirit than what holy orders typically entail.

Maskelyne was certainly facing down a host of rich associative memories—if not quite spirits or ghosts—on this lonely island, too. Eighty-four years before on St. Helena, the legendary Edmund Halley had observed a transit of the planet Mercury (when Mercury briefly passed in front of the sun). "I very accurately obtained," Halley told a 1716 meeting of the Royal Society, "with a good 24-foot telescope, the very moment in which Mercury, entering the sun's limb, seemed to touch it internally, as also that of his [i.e., Mercury's] going off."

The problem was that Mercury was so small and so far away that it required both a sizable telescope and one of the world's finest astronomers to spot and track a Mercury transit. If only Mercury were bigger and closer to earth, Halley surmised, the most vexing problem in astronomy could be solved. But, of course, there is a planet that meets this description. So, Halley realized, Venus transits were the answer. At the time of the next transit, Halley proselytized in his writings and lectures, the great nations of the world should put multiple observers at different locations across the planet. Then they could combine their Venus transit measurements to discover the angular shift the sun makes against the sky when observed from different locations on earth. Given an accurate number for the sun's angular shift, or "solar parallax," the physical distance to the sun was a schoolboy's geometry problem. However, the showstopper came when Halley calculated the date of the next Venus transit. The world would have to wait, he found, until 1761. The true dimensions of the solar system would next become visible only in the year of Halley's 105th birthday. On his deathbed in 1742, Halley knew he was still a generation away from this watershed moment in human history.

It all had started on St. Helena. The island was something close to sacred ground for men like Maskelyne who admired Halley and devoted their careers to carrying his legacy forward.

So Maskelyne drank down his failures at St. Helena—he'd also tried and failed, using a different method, to measure the distance to the star Sirius—and, together with Mason, prepared to return home.

Some might have crumpled at such a low ebb. But Maskelyne didn't even idle aboard ship. His and Mason's passage to England aboard the East India clipper *Warwick* (escorted by the 24-gun naval frigate *Terpsichore*) would be their teaching lab. Maskelyne and Mason were, after all, experts in the science that was primed to solve the greatest earthly problem known to every explorer and naval captain of the day.

Maskelyne and Mason spent their nights, dimly lit by the phosphorescent glow of the Ethiopic Sea (today's southern Atlantic), instructing *Warwick*'s officers how to measure their longitude onboard a moving ship. Maskelyne had with him a new set of nautical tables that predicted the moon's position in the sky with uncanny accuracy. Using a ship's quadrant—an iconic mariner's device with spyglass on top and wooden or metal wedge of a circle hanging beneath it—Maskelyne gave some of the first master classes on navigating by lunars. His students learned accurate shipboard measurements of the moon's angular separation from the sun (during the day) and known bright stars (at night).[29] And, equally important, Maskelyne taught the laborious calculations needed to turn these measurements into longitude. (Grinding out a single longitude value via the lunar method could take up to four hours of pencil work.) But Maskelyne and his new students proved, sometimes multiple times per day, that they could reckon their longitude to within a degree—which translated to forty nautical miles at the latitude of the English Channel.

These were the kind of results that might win the coveted Longitude Prize.

On May 14, 1762, the *Warwick* anchored at Plymouth, returning Mason to the site of his dressing down by the Royal Society. But the private humiliation he and Dixon faced the year before paled in comparison to the competitive kick in the hindquarters that Mason and Maskelyne had received in their absence.

As the astronomers offloaded their gear onto English soil, word began spreading around the docks of another ship that had anchored at Portsmouth two months before. The sloop *Merlin* had reached port on March 27, carrying onboard a watchmaker named William Harrison and the astronomer John Robison. Harrison had traveled to Jamaica and back testing his father's design of a compact, spring-wound maritime clock. If Harrison's machine succeeded as stunningly well as its promoters claimed, it could cast the entire future of lunar navigation—and, with it, many astronomers' well-funded careers—to the westerly winds.

FLYING BRIDGES

TOBOLSK, SIBERIA
June 6, 1761

The cloud bank to the east glowed red. Jean-Baptiste Chappe d'Auteroche had been living at his mountaintop observatory, avoiding contact with the superstitious townsfolk and instead looking to the skies. His lot, he'd surmised, was not to make inroads with the locals but rather, as a fellow French philosophe put it, to make "a communication of flying bridges, as it were, that reunite one continent with another and pursue all the tracks of the Sun."[1]

Three days before, Chappe had pointed his nineteen-foot telescope at a solar eclipse, recording in his logbook the exact moment when the eclipse ended. His pendulum clock—which he'd previously set to noon when the sun reached its highest altitude in the sky—read 6:11 AM and 4 seconds. He'd already calculated that the same solar eclipse would be visible in St. Petersburg as well. So when he later returned through Russia's capital city en route to Paris, Chappe could then compare notes with observers there. Since the eclipse ended at the same instant, whether seen from Tobolsk or St. Petersburg, the difference in time between these geographically separated measurements was exactly the difference in longitude between the locales. Chappe derived that Tobolsk was 65.8490 degrees east of the National Observatory in Paris. (Today, Chappe's longitude would be written as 68.1862 degrees east of the prime meridian, the British Royal Observatory in Greenwich, England. By either standard, Chappe's error was an impressive 0.0719 degrees or 4.3 arc minutes—translating to 3 miles at Tobolsk's latitude.)

The night before the transit, all looked calm. "The sky was clear," Chappe recalled. "The sun sunk below the horizon free from all vapors. The mild glimmering of the twilight and the perfect stillness of the universe completed my satisfaction and added to the serenity of my mind."

By morning, however, the 4:30 sunrise had brought a dark veil. Clouds loitered. As the increasingly cloudy and sleepless night progressed, Chappe paced the observatory floor. His assistants, whom Chappe had woken earlier in the night, left their master alone—knowing they'd only be needed if clear skies returned. "I found myself relieved by their absence," Chappe wrote.

Soon after dawn, Chappe heard a commotion outside. Tobolsk's governor, the local archbishop, and some nobles had assembled at the new observatory to take in the heavenly spectacle.[2] The first light of day shone upon the French visitor whose anxiety grew with each troubled glance at the clouded-over sky.

"The idea of returning to France, after a fruitless voyage, of having exposed myself in vain to a variety of dangers," he recalled, "[with every] expectation of success, which I was now deprived of by a cloud ... threw me into such a situation as can only be felt."

Chappe had instructed his assistants to set up a tent outside the observatory with the secondary telescope. The arrangement provided all they'd need to view the transit—but still permit Chappe to perform his own delicate observations with the privacy he demanded.

As the dawn's blush gave way to early morning light, an easterly wind peeled back the top layers obscuring the sun. And with the increasing transparency, the mood both inside the observatory and in the nearby tent lightened. "The clouds began to exhibit a whitish color, which grew brighter at every instant," Chappe wrote. "A pleasing satisfaction diffused itself through all my frame and inspired me with a new kind of life."

To everyone's pleasant surprise, the residents of Tobolsk—so vocal in their opposition to the Frenchman's entourage weeks before—had shut themselves up in their houses and churches, some fearing God's imminent wrath. Today, the armed guards assigned to protect Chappe proved an unnecessary precaution. Chappe instead enlisted their help in moving his nineteen-foot telescope out onto the lawn.

Doomsday had been postponed. Instead, the morning of June 6 brought a clear patch through which Chappe could view unobstructed the first hints of a tiny black sphere piercing the sun's sacrosanct disk. The assembled crowd in the nearby tent now had something to see.

Henceforth no excuses remained. Heartbeats quickened as Chappe cued his interpreter inside the observatory to shout out every second of every minute on the pendulum clock. A stream of numbers cut the hush with metronomic quickness. "Cinqante-cinq minutes et un . . . deux . . . trois . . . quatre . . ."

The seven o'clock hour approached as Chappe adjusted his telescope. Like a smooth, circular pebble descending into a thick fluid, Venus began crossing the solar limb. Eighteen minutes after Venus first excited the assembled crowd with its initial appearance, the first crucial moment of its solar transit approached—when the sun had enveloped the entirety of Venus's shadow. No words the interpreter had ever said meant more to Chappe than the sequence of numbers he shouted through the observatory door. "Vingt-quatre . . . vingt-cinq . . . vingt-six . . ."

At 7:00 AM and 28 seconds, Chappe recorded in his logbook the moment of internal contact between Venus and the sun. "I . . . felt an inward persuasion of the accuracy of my process," Chappe recorded. "Pleasures of the like nature may sometimes be experienced. But at this instant, I truly enjoyed that of my observation and was delighted with the hopes of its being still useful to posterity when I had quitted this life."

Many other observers of the 1761 transit reported difficulty recognizing the instant of interior contact, elongating the measure of a moment into a guessing game extended over tens of seconds. This surprise phenomenon, ruining many otherwise useful transit observations, results from an optical illusion that makes the sun's limb briefly pucker inward with a plasticity that appears to connect it with Venus's distorted disk. Chappe, on the other hand, reports no problems with what was later dubbed the "black drop effect."

The aromatic musk of Russian tea—hints of honey and Spanish pepper—spiced the air of this increasingly beneficent morning. Chappe had refused dinner the night before, and the twelve or more hours since his last meal would have left the explorer with little more than adrenaline to fuel him. The city's nobles, gathered around the nearby tent, provided a counterpoint. Spirits like bilberry wine and quas—a commonplace Russian drink made of fermented meal and malt[3]—likely blurred these momentous few hours into pleasantry while delicacies like caviar and roasted quail tempted hungry men in Chappe's party to join in. ("All these [Siberian game] birds," Chappe grumbled, "have a disagreeable fishy taste.")

The sun continued rising, and the clouds continued to clear. Around 10:00 AM, Venus had reached the halfway point in its transsolar journey. At the exact median, Chappe tended to the second smaller telescope to make a different kind of measurement. As an independent check against the transit's time records, he also measured the separation between the nearest edge of Venus and the sun's enveloping arc. His ten-foot telescope, in less demand after drink and disinterest had peeled away some of the observatory's guests, made its 3,000-mile journey to yield one crucial number of angular distance. Inside the smaller telescope's eyepiece a translucent set of hash marks provided the ruler that yielded 6 arc minutes and 2 arc seconds of angular separation between planet and star at the transit's halfway point. And to check his check, Chappe also measured out the entire diameter of the sun: just over one-half of a degree—31 arc minutes and 37 arc seconds.

Now measured as if for a new outfit, a star ascended. It gave day to the earth, as it always does. But on this day its closest watchers had reached into the beyond for their first grasp at a universe of knowable depth.

Imperial Academy of Sciences, St. Petersburg, Russia
January 8, 1762

The lecture everyone had come to see concerned a tiny planet passing in front of the sun.[4] But from all the chatter in the room, a casual observer might think the sun had gone into permanent eclipse. A fortnight before, fifty-two-year-old Elizabeth Petrovna, empress of Russia, had died.

Elizabeth had reigned for twenty years over relative domestic calm during an age of nearly nonstop European war. She was widely beloved, both at court and in the populace at large. She was, nevertheless, every bit a Russian czarina too. Her extravagances were legend: 15,000 ball gowns and thousands of pairs of shoes. And not a few lovers on the side. She also brooked no personal criticism. In his memoirs of the Siberian voyage, Chappe tells a chilling story of Elizabethan retribution against courtly ladies who had conspired against the empress—although some historians suspect the women were in fact only guilty of talking too freely of Elizabeth's amorous activities. In a riveting story that was cited and re-recounted for generations to come, including in the legendary *Life of Samuel Johnson,* Chappe wrote:

> Madame Lapouchin was one of the finest women belonging to the court of the Empress Elizabeth . . . [and] was condemned by the Empress Elizabeth to undergo the punishment of the knout. She appeared at the place of execution in a genteel undress, which contributed still to heighten her beauty. . . . One of the executioners . . . then took a kind of whip called knout,

made of a long strap of leather prepared for this purpose. He then retreated a few steps, measuring the requisite distance with a steady eye. And leaping backwards gave a stroke with the end of the whip so as to carry away a slip of skin from the neck to the bottom of the back. Then striking his feet against the ground he took his aim for applying a second blow parallel to the former. So that in a few moments all the skin of her back was cut away in small slips, most of which remained hanging to the shift. Her tongue was cut out immediately after, and she was directly banished to Siberia.[5]

Despite a brutality that matched her vanity, the empress had endeavored to make St. Petersburg a European capital of high culture—graced with her characteristic enchantment of French culture, in particular. So although some Russian academy members had resisted supporting Chappe's Siberian voyage, the academy nevertheless offered the visiting Frenchman the opportunity to discuss his work and findings.

Still, one might forgive Chappe's audience for a level of distraction higher than any scholarly norm. The late empress had also left behind an empire apprehensive of its future. Now holding the scepter of power in Elizabeth's wake was her reviled nephew Peter—"the most imbecile prince that ever ascended the throne of a vast empire," as one contemporary journalist described his public reputation.[6] Elizabeth's armies— allied with France and Austria against Britain and Prussia—had beaten back the Prussian army to the point that total victory was thought to be close at hand. But Peter so worshiped the Prussian commander Frederick the Great that rumors were spreading that the new Russian emperor wanted to completely withdraw Russia from the war, consequences be damned.

Peter III was preparing to commit his nation—and the Russian-French-Austrian alliance—to a mass act of strategic suicide. Not out of any inviolable principle or budding pacifism either. The newly crowned

Russian leader, it was said, liked nothing more than playing with his toy soldiers. He was 33.

Chappe looked out at the audience, a gathering of some of the empire's greatest minds. The eminent visitor's lecture, later printed in French in St. Petersburg, gloried in the accomplishments of his mission to Tobolsk.

"Astronomers [have] waited for over a century for the transit of Venus across the sun's disc," Chappe began. "This phenomenon that seized their fancies seemed all the more important in its promise of a whole new day for astronomy."[7]

The guest speaker told his audience that the Venus transit mission from which he'd returned two months previous reflected brilliantly on the larger Russian court. He strategically left out any mention of the Russian astronomers whom the Russian Academy of Sciences had sent into the remote hinterlands for their own Venus transit observations that might compete with Chappe's. One expedition was, so far as can be documented today, never heard from again. But another, led by the astronomer Stephan Rumovsky, did ultimately return with Venus transit data from the Lake Baikal region.[8] Rumovsky's work and findings go unmentioned in Chappe's talk, however.

Instead, Chappe discussed his own data and paid genuflecting homage to the empress who had approved of his own Venus transit voyage to Siberia. Chappe told his audience, indeed, that the greatest political leaders reign over societies where the arts and sciences flourish. And without drawing too much attention to the current occupant of the Russian throne, Chappe heaped praise instead on Peter III's grandfather Peter the Great, who, Chappe said, "knew that the sciences and arts are also closely related to their glory."

Chappe's words resounded through the very same palace that Peter the Great had given over to the creation of a library and academy. To further the academy's mission, the dearly departed Elizabeth had in her dying days offered Chappe the position of imperial astronomer of

Russia. (Or so Chappe claimed, at least. His mentioning such an un-confirmed job offer from the late empress might also have served as a none too subtle dig at the France-hating Russian polymath astronomer Mikhail Lomonosov, who by all rights should probably have been im-perial astronomer.) The post of Russia's chief astronomer had lain va-cant since 1747, when Chappe's countryman Joseph-Nicolas Delisle had abandoned the office that he'd held for twenty-two years and returned to his native Paris.

Even if the job offer were for real, Chappe evidently had other designs on his future than to remain in St. Petersburg. A second Venus transit was seven and a half years away. And this time, according to the predic-tions concerning the regions of the earth that would see the whole tran-sit, Russia held little valued real estate. In 1769, locations in or near the Pacific Ocean would instead be some of the world's best destinations.[9] Serving the Russian court would not be nearly as strategic a choice to collect some of the best data on earth from the 1769 transit. Jockeying for top position within the Royal Academy of Sciences in Paris at least afforded him the chance.

Chappe also subsequently admitted an underlying prejudice—one that his own mission's financial independence from Mother Russia en-abled him to hold. There was, he thought, something about the nation as a whole—perhaps the Russian weather—that made people dumb. "The state of the arts and sciences in Russia implies a defect," Chappe later wrote, clearly in a more critical mind. "The cause of which must be sought for, either in a want of genius particular to the nation, or in the nature of the government and the climate."[10]

Such jaw-dropping candor, of course, was absent from the St. Pe-tersburg lecture hall where Chappe held forth. Rather, he paid his homages and thanked his gracious hosts. The "graces and virtues" of the late Elizabeth, he told the crowd, "cultivate the sciences with great success. . . . What a good omen for the progress of this empire! I see them [those graces and virtues] contributing to the glory of the nation and of Peter III."

In July, when Chappe would be in transit to Paris, conspirators plotting with Peter III's German-born wife (and second cousin) Sophie would overthrow her husband and crown Peter's bride the new autocrat of Russia. Peter III's captors thereafter relieved their prisoner of his pulse. And the preternaturally canny foreign empress—who'd learned the Russian language and converted to the Orthodox religion—soon cast off her widow's weeds to rule Russia under her assumed name, Catherine the Great.

THE ROYAL OBSERVATORY AT PARIS
August–November 1762

Nearly a century before the Venus transit voyages, King Louis XIV established the Royal Academy of Sciences in Paris. At the time, the British had established a new Royal Society, placing England's royal imprimatur on scientific research, at a time when science was gaining economic and political relevance. France did not chance the British taking the lead. Within a year of the academy's founding, Louis's ambitious finance minister, Jean-Baptiste Colbert, set to building it a home—a royal observatory on the southern outskirts of Paris. (From the academy's inception, the navigation problem was central to its existence. Two-thirds of the initial membership of the Royal Academy in 1666 studied either astronomy or geometry.)[11]

It was to Colbert's stone fortress that Chappe returned in August 1762, after his twenty-one-month voyage into Siberia and back again. Those same ramparts had also become the world's clearinghouse for Venus transit data, and Chappe's colleague within the Academy of Sciences, Jérôme Lalande, was positioning himself as one of the world's leading experts on the Venus transit.

A dwarfish man with a bad back and worse eyesight, Lalande was ill-equipped to make physically taxing expeditions.[12] Instead, he made his name as a calculator and a commentator.

On November 18, Lalande wrote a letter to his British counterpart, Nevil Maskelyne, which the *Proceedings of the Royal Society* excerpted. In it, Lalande shared the first round of results the French had acquired from several Venus transit expeditions. He had many to choose from. In Germany, a mathematician reported observations he'd made from the orangery of Schwetzingen Castle near Heidelberg; two farmers near Dresden (in independent observations) made their own amateur measurements of the event; a doctor and a Jesuit monk near Nürnberg and Tyrnau, each transmitted their observations too. In Sweden, the country's queen and prince joined a throng of prominent spectators crowding into the observatory at Stockholm—so crowding the astronomers that they couldn't even see their clock. In France, two Catholic priors in Rouen observed the transit from the astronomical observatory at their monastery; telescopes at an abbey in Paris and episcopal palace near Montpellier tracked the planet's motion across the solar disk. Astronomers in England took to the fields, farms, and observatories across the country, while the Duke of York summoned a London instrument maker to show him the famed transit. The duke's observing party never saw the event due to inclement weather.

Ultimately 120 Venus transit teams from around the world would report two key sets of numbers from the 1761 event: the precise latitude and longitude of their observing station (along with the methods they used to determine their coordinates on the globe) and the time that Venus took to cross the sun. Some also tracked the precise path Venus took across the sun, enabling an independent check of the solar parallax derived from the timing measurements. Some locations, such as Friar Hell's observatory in Vienna and Mason and Dixon's in Cape Town, could only observe the latter half of the transit. In such cases, the observers reported the exact moment the transit officially ended, when an exiting Venus's outer edge first touched the sun's inner edge.

The data, in aggregate, effectively describe a triangle, with one vertex at the center of the sun, another at the center of the earth, and the last vertex at an outer edge of the earth's disk as seen from the sun. (Precisely how one gets from timing and latitude/longitude data to this triangle can be found in the Technical Appendix at the end of the book.) The triangle is superthin: the angle between the triangle's two long arms is so small that if it were a wedge of pie, one could serve up another 144,000 similarly slim pieces before cutting into a second pie.

That angle is called the "solar parallax." It's close to the slight angular shift of the sun's position in the sky when observations from the equator are compared to observations from the North Pole. (The angular shift would be exactly the same as the solar parallax but for the fact that the earth also spins on an axis that's tilted compared to its orbital plane. In any event, parallax can be readily demonstrated by holding up a pencil at arm's length and closing the left eye and then the right eye. The left-eye/right-eye parallax is approximately 1,400 times bigger than the solar parallax.)[13]

Lalande found, as he reported to Maskelyne, that all the French expeditions taken in total yielded a solar parallax of 9.55 arc seconds—translating to 85.3 million miles distance from the sun. Lalande added that a fellow French astronomer, Alexandre Guy Pingré, had recently returned from observing the Venus transit on the Indian Ocean island of Rodrigues. And Pingré had derived the same value for the solar parallax.

Later analysis would reveal that Pingré had in fact arrived at a substantially different answer: 10.6 arc seconds or 76.8 million miles.[14] But Pingré's putative confirmation of Lalande's result was all Lalande needed to conclude that France was in the right. Maskelyne had previously reported his first analysis of the solar parallax: 8.6 arc seconds. But Lalande dismissed the English number as the result of flawed data from Swedish observations of the transit. (Lalande was wrong. As present-day observations reveal, the correct answer that everyone was chasing was 8.794 arc seconds.)

Lalande further marveled at reports coming in from the field that some observers had difficulty pinpointing the exact moment when Venus's silhouette touched the sun's edge. The "black drop effect" had blurred the precision timing of this crucial instant by seconds or minutes. On the other hand, Lalande had witnessed the latter part of the Venus transit from Paris and, as he put it, "I was not uncertain so much as a single second."[15]

On the other side of the channel, Maskelyne took the widely varying solar parallax results as a sign of polluted data. "I'm afraid we must wait till the next transit in 1769," Maskelyne reported to the Royal Society, " . . . to do justice to Dr. Halley's noble proposal and to settle, with the last and greatest degree of exactness, that curious and nice element in astronomy, the sun's parallax, and thence determine the true distance of all the planets from the sun and from each other."[16]

The world would have to try again.

Chapter 4

THE MIGHTY DIMENSIONS

Londonz
April 15, 1763

War was over.

England and Prussia had led a loose-knit coalition of allies to battle-field triumphs in North America, Europe, India, and western Africa—and to naval victories in the Atlantic, North Sea, Mediterranean, and Caribbean. "The flame of war," one contemporary chronicler wrote, had been kindled "in every quarter of the world, and which afterwards raged . . . with a destructive and unrelenting fury, beyond the example of former times."[1]

Spain, Portugal, and France had surrendered to Britain in February 1763, while at the same time Austria ratified a separate treaty with Prussia. Russia had withdrawn from the conflict due to monarchical in-transigence, while "war's end" in North America meant sowing the seeds of bloody conflict for generations to come with indigenous tribes across the continent. In the age of Voltaire, lunar longitudes, and rococo, even peace was complicated.

King Louis XV of France was willing to accept all but one of the pro-visions of the Treaty of Paris, which ended the Seven Years' War: Britain insisted that France level its towering coastal forts at Dunkirk. England feared that a heavily armed Dunkirk—from whose citadel one could practically see the mouth of the Thames—would be a perfect staging ground for French mischief. The French king brooded.[2]

Ultimately, having forfeited the perfect launching point for a military invasion of Britain, Louis began planning in secret for a military invasion of Britain. Heading up Louis's entourage of personal spies in England was a diplomat and soldier named Charles Geneviève Louis Auguste André Timothée d'Éon de Beaumont. D'Éon was a small and slippery man hiding behind a billboard of a name. His reputation for disguise was legend. Rumors circulated that only a few years before, d'Éon had penetrated Russian empress Elizabeth's inner circle in St. Petersburg by posing as one "Mademoiselle Lia de Beaumont," a "niece" of a distinguished British tourist (who was also a French spy).[3]

In March 1763, King Louis had sent d'Éon to London to secretly scout out the English countryside for enemy troop garrisons and possible landing sites for a French strike force. The king kept nearly all of his advisers ignorant of the undercover mission. Louis would later write down d'Éon's sub-rosa orders, to make "reconnaissances . . . in England, be it on the coasts, be it in the interior of the country . . . [and to] keep this affair strictly secret and . . . never mention anything of it to any living person, not even to my ministers."[4]

On Friday, April 15, d'Éon met in London for a private dinner (the noontime meal) with the visiting astronomer Jérôme Lalande. Lalande had been traveling around the greater London area for four weeks. When town criers announced the armistice on March 22, British patriots on the streets near Whitehall shouted down the visiting Frenchman. "You are stupid," they taunted Lalande, "like the peace!"[5]

Other than this pugnacious encounter, Lalande enjoyed something close to VIP status in London. During his stay, Lalande had been meeting mostly with Englishmen—instrument makers, scientists, authors, and businessmen—and was working toward his own ultimate goal: nomination to the British Royal Society.

Lalande and d'Éon had first met at a dinner party at the French ambassador's residence in town.[6] In the interim, Lalande had learned that his personally funded voyage to England was about to take a turn toward the national interest.

King George III had recently given his royal assent to "An Act [of Parliament] for the Encouragement of John Harrison to Publish and Make Known His Invention of a Machine, or Watch, for the Discovery of Longitude at Sea."[7] According to the terms of the act, the watchmaker John Harrison was sometime soon going to disclose the secrets behind his revolutionary new nautical chronometer—a watch that was said to keep time so well that over a recent eighty-one-day sea trial to Jamaica, it allegedly lost only 51 seconds. By comparison, a good standard-issue naval clock might gain or lose four minutes per day.[8] This slippage meant that any navigator who relied on it for longitude could be off in his estimate of a ship's position by fifty miles or more. Rocky shores could find safe harbor in such fuzzy numbers, damning the most powerful gunboats in the world to ignominious death. Here beat the heart of the beast that swallowed ships and legions of sailors whole.

But if Harrison was right, the game was about to change. If Harrison's astounding feat could be replicated—and if his watch could be affordably duplicated—the centuries-long longitude quest might well be over. One of these new chronometers could keep London or Paris time onboard any ship. And comparing it to the local time anywhere around the globe, the ship's longitude could be immediately and easily calculated. As a result, Lalande and his fellow astronomers might also lose royal patronage and attention, becoming as outmoded as a regiment of pikemen in an age of musketeers. And England and France, and not a few other colonial nations besides, would be racing to sea and across the globe armed with a technology that would surely reshape the planet.

Then, five days before his dinner with d'Éon, Lalande learned of a scientific mission to England to pry out the secrets of John Harrison's phenomenal machine. Moles in the British government had assured the French ambassador that Harrison would be making his disclosures in public—or at least in a forum where well-connected Frenchmen could gain access. The visiting French horologists would thus be on hand to uncover, as Lalande wrote in his diary, "the secrets of longitude from Harrison."[9] And Lalande—fluent in English as none of the other French experts were—would serve as their unofficial liaison.

D'Éon lunched with Lalande that Friday, and the two spent the afternoon west of town. Their initial stop, Hyde Park, was a popular locale for Londoners to "take the air." The park's grounds gave a breath of Elysium to a choking city. More than a few locals knew this too. One London wit said Hyde Park was bathed in beauty by the fashionable women who strolled its grounds—just as it was drenched in dust kicked up by the horses who carted them around.[10] Lalande recalled the park being "extremely pleasant for coach drives and riding on horseback."[11]

Their ultimate destination was Kensington Palace, a longtime residence of the recently deceased King George II.[12]

The astronomer and the spy walked down Kensington's "beautiful lawn, which surrounds the lake on the London side," Lalande recorded. Inside, they admired the portraits in the palace's gallery and the ornate, marble-pillared and gilt-statued "cube room." Although Lalande was only an unwitting accomplice, d'Éon continued to carry out his orders to reconnoiter strategic London locations, with an eye toward a day when France might turn the tables on its perennial enemy across the English Channel.

While d'Éon never dropped his cover, Lalande had loose lips. Lalande's proximity to the daily developments in, as Lalande put it, "the discovery of Harrison's secrets" surely caught the spy's ear. D'Éon continued meeting with Lalande for another eight dinner dates and assorted other diplomatic engagements.

Two days before the Kensington Palace outing, Lalande noted in his diary how he'd learned that commissioners carrying out the Act of Parliament would reward Harrison for his work and open the door for the French watchmakers to learn Harrison's tricks. Harrison was unhappy with what the commissioners expected of him. He felt that British officials didn't appreciate the potentially grave consequences that a public disclosure of his clock's design could bring about.

Harrison would be swatting away the French flies buzzing around him for months to come. And for good reason too. Not all were as incapable at espionage as the petit, self-aggrandizing, and conspicuous French philosophe who never quite understood the excitement over marine chronometers in the first place.

London
April 22, 1763

Lalande had now been to Parliament, where he'd met King George III. (Another dinner with d'Éon promptly followed.) And the next Friday, Lalande called on his colleague Nevil Maskelyne, who lived in a fashionable neighborhood of London, near Hanover Square. Wedding bells in Maskelyne's Mayfair region of town often rang for, as one chronicler put it, "swell marriages . . . [for] many a belle of the London Season" conducted at St. George's church down the street.[13]

And scarce were there two less romantic people in the city—well disposed to curse those damned bells in two-part harmony—than these gentlemen of mathematical and celestial calling. Table talk, at such a propitious moment for the host, could hardly have avoided the obvious. Maskelyne had been hard at work for nearly a year on a new book that was now just appearing in bookseller's stalls throughout the city. Maskelyne's *British Mariner's Guide Containing Complete and Easy Instructions for the Discovery of the Longitude at Sea* worked exactly as advertised. It compiled tables, instructions, and lessons Maskelyne learned in measuring lunar longitudes during his 1761 Venus transit expedition. Maskelyne began his tome with the promise, "I . . . can from such experience venture to answer that this method carried into practice will (without disparagement to the labor and inventions of others) bring the longitude to great nearness."[14]

Maskelyne and Lalande were men of the same mind. "I do not see why the longitude might not be as universally found at sea by this method as the latitude is at present," Maskelyne said. As Lalande wrote in his own French almanac, *Connoissance des mouvemens célestes pour l'année 1762*, the lunar longitudes method "every day becomes easier and more accurate. Oh that we can make it as widespread and familiar among navigators! The benefits they derive will be immense."[15]

As far as these two astronomers were concerned, the solution to the greatest problem of the day was at hand. What remained for the solution to be handed over to the world was a streamlining of their method and an instructional curriculum that taught their art.

Maskelyne and Lalande had long communicated from afar about astronomy, lunar longitudes, and the Venus transits. Here at last, as the two men broke bread together, they could collaborate in person.

Of the two men, Maskelyne would be chief evangelist. Lalande had too many varied interests to devote the kind of singular focus his British counterpart possessed. Even during his British visit, for instance, Lalande was also reading over proofs of a forthcoming book he'd written on the art of tanning chamois leather.[16]

Maskelyne, on the other hand, had in one book practically defined his career. The guide lays out in twelve pages of exquisite detail the four measurements his method required to discover a ship's longitude at sea: The angle between the horizon and the sun (a.k.a. the "solar altitude"), the lunar altitude, the angle separating sun and moon, and the time of the measurement. (If done at night, a bright reference star is used in the place of the sun.) Another sixteen tables over thirty-four pages laid out every finicky detail a navigator needed to know to determine his ship's longitude accurately, from the height of the ship's deck above the surface of the water to the exact predicted location of the moon in the sky.

Predicting the moon's precise position a year or more in advance had for generations kept longitude outside human grasp, a mathematical and geopolitical will-o'-the-wisp. In its twenty-eight-day orbit around the earth, the moon advances almost 13 degrees through the sky every twenty-four hours. Its motion against the background stars is almost like the steady advance of an hour hand on a clock as big as the celestial sphere. If it were that simple, of course, determining longitude would never have been any more difficult than latitude. But exact prediction of the moon's motion is in fact very complex, because the sun slightly distorts the moon's path—over an elliptical orbit that forever wobbles like a spinning top owing to its 5-degree inclination compared to the orbit of the earth around the sun.

It took a German mapmaker in 1755, Tobias Mayer, to formulate the complex framework of equations that forecast the hourly positions of the moon every day and night for months or years ahead of time, an act of hardscrabble genius that earned him £3,000 from the British Board of Longitude. (Mayer had died in 1762, however. So his widow came to England to collect the reward—and met with Lalande during her English visit.)

Maskelyne tested Mayer's lunar theory on Maskelyne's Venus transit voyage and found that a couple of minutes of measuring and four hours of calculating provided reliable longitudes "always within about a degree and generally within half a degree."[17] But Maskelyne also discovered that most of his numbers could be crunched ahead of time and printed up in tables no more complex than hackney coach schedules. Four hours of labor could be thus reduced to thirty minutes. Maskelyne's lunar longitude method, one contemporary reviewer noted, "may seem a little troublesome . . . but we are informed that a very little practice will render it easy and familiar."[18]

At the time of his lunch with Lalande, Maskelyne had every cause to suspect the *Mariner's Guide* would be setting a new standard for navigation—enabling both commercial and military ships around the world to keep their course and taming some of the fiercest savageries of the open seas.

The *Mariner's Guide* also planted a flag. As a convenience to him and his colleagues at the Royal Observatory, Maskelyne centered the *Mariner's Guide*'s longitude tables at Greenwich. A subsequent set of annually published tables that Maskelyne would supervise (*The Nautical Almanac*) would use the same convention. Partly because he couldn't top them, Lalande would reprint Maskelyne's longitude tables verbatim in French navigational almanacs from 1772 onward.[19] For almost a century, three-quarters of the shipping tonnage around the world used charts based on Maskelyne's standard.[20] And these two friends, brought together over a few lunches in the spring of 1763, would ultimately enshrine Greenwich, England, as the reference point for keeping "universal time" as well as anchoring the earth's zero-degree longitude prime meridian.

LONDON
May 8–9, 1763

For all the enthusiasm Lalande and Maskelyne exuded over lunar longitudes, some London correspondents had already picked their darling. It was fancy, slick, and full of gadgetry. Its inventor, Christopher Irwin, had a knack for public relations. He'd garnered the endorsement of a naval war hero newly elevated to the peerage after his viscount brother had died in action at a celebrated battle in the American colonies. The story was sexy.

The *British Palladium* reported that Irwin's new device left "no doubt of the Longitude's being discovered and settled to a very useful nearness." The *Monthly Chronologer* relayed how both Prince Edward and the king's "mathematical teacher" had tried the gadget out. "This will do!" the latter reportedly cried. "This will do!" *The Annual Register* noted, "Navigators are, for the future, to consider [its] invention . . . as one of the greatest benefits that can possibly accrue to their science."[21]

The invention was called the "marine chair"—a gimbaled and counterweighted seat designed to hold its sitter still on a rocking and swaying boat. A telescope was affixed to the chair. Using the device, a sitter could then view the steadiest-ticking celestial clock in the solar system, the moons of Jupiter. With the reliable Jovian satellites serving as chronometer, longitudes might now be available down to a record-breaking one-third of a degree.

The London instrument maker Jeremiah Sisson had already shown Lalande one prototype marine chair. But now, two French horologists—those same peering eyes hoping to pry out the secrets of John Harrison's watch—had arrived in town. Sisson was ready.

Sisson, who was modifying some of Irwin's marine chair designs, pulled out three models for close inspection. Sisson's shop—on The Strand, amid the clang and din of central London—crackled with the electricity of opportunity.

Though very talented at his craft, Sisson had a flighty attention span that left the dilettantish Lalande ill at ease. The craftsman had of late been pawning his handiwork for easy cash. His visitors were marked for the full-on pitch.

Out came a chair "mounted on a suspension with four pivots and two boots." Sisson showcased a second seat, held high on a "knee joint from which a 6 foot pendulum hung, with the weight at the bottom in water." Finally the French horologists viewed an eight-by-four-foot monstrosity "made up of two circles each having two pivots, but whose directions cross."

Sisson, Lalande recorded, "is obliged to work in great haste, and so achieves nothing worthwhile. But nevertheless there is no one who has as much ability as he."[22]

Lalande's two countrymen—the clockmaker Ferdinand Berthoud and the mathematician Charles Étienne Louis Camus—no doubt hoped for a similar welcoming the next day when they called on John Harrison at his nearby house in Red Lion Square.

More serious and steadfast in his persistence, Harrison was a man of defiant will who, for starters, had once dared to doubt the greatest scientific genius in history. In 1725 Sir Isaac Newton had informed the British Admiralty that two, and only two, practicable methods for finding longitude at sea existed: lunar longitudes and longitudes via the moons of Jupiter.[23]

Moreover, latitude at sea had always come courtesy of measurements of the sun and stars. Expecting manmade springs and gearwheels to solve longitude was almost an affront to the heavens. As the French mathematician Jean-Baptiste Morin put it, "It would be folly to undertake it. . . . I don't know if the Devil himself could do it."[24]

To both Lalande and Maskelyne—commanding increasing authority in their respective countries—Newton and Morin had stated what to their generation of natural philosophers had become self-evident. Marine chronometers were a pointless dream, a wild and distracted sprite. Longitude's future was with the moon.

But Harrison simply would not go away. A generation before, in 1735, Harrison completed his first sea clock—a portmanteau-size, seventy-two-pound, gear-shafted dial box that flexed long arms of spring-propelled brass spheres. The next year the Royal Society sent Harrison and his temporal dynamo to Lisbon to test it out. Absurd though the chronometer looked, it yielded impressive results on the sea journey there and back. For starters, when Harrison's ship first returned to the channel, Harrison accurately informed the captain which English islands were coming into view. The navigation officers had instead located the ship sixty-eight miles east of where it actually was.

The English Board of Longitude awarded Harrison £500 to develop a better chronometer. So Harrison completed his second machine in

1739, making a timekeeper bigger and heavier than the first. It still wasn't reliable or rugged enough, however. In the next twenty years, he tinkered and fussed and ultimately made two breakthrough designs that better compensated for jostling seas and temperature extremes.[25] Harrison finished his third marine chronometer in 1759, another un-wieldy box. Then, just one year later, he was ready to unveil his true masterpiece—a palm-size, three-pound pocket watch with all the ocean readiness of his previous chronometers.

This fourth Harrison design, presented to the Board of Longitude in July 1760, was the stroke of genius that had set French spies to knocking.

Miscommunications in Harrison's 1761 transatlantic trial of his sea watch left commissioners from the Board of Longitude unsatisfied, con-cluding that another transatlantic trial was needed. And so, in the spring of 1763, the master craftsman awaited the final opportunity to vindicate himself.

Harrison had in fact shown his masterpiece watch to Lalande three Fridays before, on April 22. But that was before Berthoud and Camus had arrived in England. Lalande knew nothing about watchmaking. La-lande was no more able to appreciate Harrison's meticulous handiwork as Harrison was able to recognize the enduring brilliance of the recent breakthroughs in lunar longitudes.

So Lalande, Berthoud, and Camus called on Harrison on Monday, May 9. Harrison obliged and opened his home to his nation's former enemies, setting out for their analysis and perusal his first three sea clocks.

Berthoud, Lalande noted, "found these pieces very beautiful, very clever, very well executed. And though the regularity of [Harrison's] watch was quite difficult for him to believe, he was even more impatient to see it after seeing the three clocks."[26]

Lalande translated his countrymen's requests to inspect Harrison's pièce de résistance. This is where the tour stopped. Parliament and the

Board of Longitude may have been hopelessly naive in mandating open inspections of Harrison's handiwork. The craftsman himself, however, was a little more cagey.

Berthoud was left to gather what intelligence he could from Harrison's first three chronometers—and from Lalande's inexpert recall of Harrison's marine watch.

Nevertheless, Berthoud had learned enough to return to Paris and reconfigure his whole clock-making enterprise. Four of Harrison's innovations soon found their way into Berthoud's Horloge Marine 2. And Berthoud further turned out two marine watches, building on the knowledge and inspiration Harrison had imparted.[27]

BRIDGETOWN, BARBADOS
October 1763–August 1764

Charles Green had everything now before him. After Charles Mason departed for the 1761 Venus transit mission, Green assumed Mason's role as second in command at the Royal Greenwich Observatory. After Green's boss—James Bradley, Astronomer Royal—died in July 1762, the next Astronomer Royal was sickly and spent most of his time at home in Oxford. Twenty-nine-year-old Charles Green was left to handle most of the duties himself. The young man had, in effect, taken the reins of arguably the most prestigious astronomical job in the world.

In September 1763, with rising star Nevil Maskelyne, Green set sail for Bridgetown, Barbados, on a new transatlantic test of the top contending longitude technologies. The original inventor of the marine chairs, Christopher Irwin, had installed two working prototypes onboard their ship, the HMS *Princess Louisa*. During the voyage, Green and Maskelyne tried out these "marine machines" in a real oceanic setting. The tests proved a disaster. The astronomers each attempted on more than one night to sight Jupiter and its moons through the chair's telescope, only to find that the planet wobbled and zipped around the scope's field of view. Close monitoring of the planet and its moons was impossible. The creaking contraption could not compensate for the unsteady sea—even on a calm night. The chair, Maskelyne wrote in a private letter to his brother Edmund, "proves a mere bauble, not in the least useful for the purpose intended."[28]

Green and Maskelyne also took multiple lunar longitudes using both Maskelyne's mahogany quadrant and a brass sextant on loan from the Board of Longitude. Their final longitude, just before making landfall off the coast of Barbados (thus making it independently verifiable), was a half degree off the true value. Combined with the lunars Maskelyne took in his Venus transit voyages, he now thought he had all the proof he needed to secure the Longitude Prize.

Still, the Board of Longitude had sent Green and Maskelyne to Barbados primarily to provide precise, independently derived times and longitudes that would enable testing of John Harrison's marine chronometer, which would be arriving on a separate ship.

To Maskelyne, Bridgetown's luscious tropical setting simply meant more cloudless nights for practicing his science. "This country is much better adapted for celestial observations than England, the air being generally much purer & serener, insomuch that for this month past I have miss'd scarce any observations that occur'd," Maskelyne wrote to his brother.[29]

The welcoming, aquamarine waters of Barbados bathed the travelers' afternoons and evenings in the sun's reflected glow. The seaside Fort Willoughby provided an open-air observatory whose warm and gentle evening breezes gave English astronomers—accustomed to night shivers huddled over a telescope—a new, tropical standard of comfort. But those same ocean breezes carried so much moisture, in fact, that exposed steel parts or instruments could rust overnight. Copper spoons portioned out the fresh cane sugar for their tea.[30]

December's cooling retained the warmth of a pleasant Hampshire summer, and colors of the Yuletide season—snowy whites and deep scarlets—blossomed on wild potato vines and Christmas bushes. Maskelyne and Green set to moving inland and sited a new observatory at the foot of Constitution Hill, 750 yards from their beachside Fort Willoughby accommodations. The eight-by-twelve-foot wooden shed was even tinier than the island's slave shanties but concealed an outsize purpose. Here the competing technologies vying to reshape the entire world would feel the cold hand of the scientific method.

Cold cash was on the line too. A prize of up to £20,000 (more than $5 million today) awaited Harrison or Maskelyne if either the marine chronometer or the *British Mariner's Guide* could definitively fulfill the terms of the 1714 Longitude Act. The two key criteria were reliable longitudes accurate down to one-half of a degree and a method that was "practicable and useful at sea."

At 1:39 AM on January 7, 1764, Maskelyne recorded the exact time of what would be sixteen immersions or emersions of Jupiter's moons over the coming eight months at the new location.[31] Although not even the finest of the pitiful marine chairs could make Jovian measurements possible onboard a ship, on land Jupiter was still the quickest and easiest route to precision longitudes.

Celestially derived longitudes may have been widely considered the only serious methods around, but the rapidly advancing watchmaking technology knew no such human prejudices. A showdown was coming. And when the HMS *Tartar* first appeared in Bridgetown's Carlisle Bay on May 13, the Caribbean winds fanned the smoldering flame.

John Harrison's son William, onboard the *Tartar* to help test the marine watch in the stead of his elderly father, had barely disembarked the ship before he'd learned of Maskelyne boasting that the prize was his. The Board of Longitude, Harrison recorded in his journal, had allegedly all but assured Maskelyne he'd already won.[32] Harrison's stunning marine chronometer longitudes on his voyage to Barbados—accurate down to one-sixth of a degree—now seemed irrelevant.[33] Was true scientific proof to be honored anywhere inside Maskelyne and Green's slat-roofed shack?

Harrison was livid. At dawn the next morning, Harrison and Sir John Lindsay, captain of the *Tartar*, appeared outside the observatory. Besmirched honor of this degree might, in a dispute between peers of the realm, call for a pair of pistols at this early hour. As it was, dueling words sufficed. The prize money looming over everyone, Harrison said, made Maskelyne unfit to provide independent verification of the marine watch's longitudes.

Lindsay brokered an accord in which Green and Maskelyne alternated the days they would be checking the marine watch against the exact solar time—a crucial element of Harrison's claim to the prize. The captain and three other witnesses stayed as rooted as the palm trees throughout, a constant presence further assuring no shady results.

Sometime during the Barbados trials Green and Maskelyne too had let the tension of the moment get the better of their relationship. According to contemporary accounts of the Barbados trials, the two astronomers suffered some unspecified disagreement that ultimately killed their friendship.[34]

On June 4, Harrison and Green set sail for England, bringing with them the rarified marine watch—all too rarified, perhaps. For it was the fact that no copies of Harrison's watch had yet been made that now prevented the Board of Longitude from awarding Harrison the prize. A solution could only be "practicable," they ruled, if it could be readily copied and affordably installed on English military and mercantile ships around the world.

Meanwhile, Maskelyne remained in Barbados until August, continuing further lunar observations he'd begun in St. Helena in 1761. Arriving in London in October, Maskelyne boasted that he'd once again used lunars to measure his ship's longitude down to the same level of accuracy Harrison claimed for his chronometer. Yet the lunar tables in Maskelyne's *British Mariner's Guide* were now beginning to lose their edge. Although technically extended all the way out to 1780, every successive month added more uncertainty to the lunar predictions Maskelyne had calculated the year before. To keep his claim alive, Maskelyne would have to perform his own duplication rites too. A new almanac would be needed.

No clear winner emerged from Barbados. Yet both frontrunners had proved themselves worthy of continued royal attention and patronage. The Bridgetown standoff carried on, though the voyage itself was fading like so many sumptuous sunsets over Carlisle Bay.

BREST, FRANCE
October 1764

For France, 1763 was a cruel year. In February the Treaty of Paris effected one of the largest and richest territorial surrenders of the century. Along with Louis XV handing over Caribbean and Indian possessions to his rival across the channel, French Canada and French Louisiana east of the Mississippi had, with the flourish of an English quill, become British possessions.[35]

But Jean-Baptiste Chappe d'Auteroche had more immediate concerns on his mind. In 1763 alone, French ships had barely averted three disasters—all of which sprang from the lack of reliable longitudes at sea. Chappe reviewed these near misses for his colleagues at the Royal Academy of Sciences in Paris. One boat traveling to the city of Cayenne in South America had overshot its destination by 150 miles. It was only spared because its surprise landfall came in broad daylight, in fair weather. Another vessel traveling to the Bermudas nearly wrecked when its longitude estimates proved wrong by 100 miles. Tall island landmarks that enabled quick course correction proved its salvation. And a third voyage, with one of the finest navigators in France at the helm, flirted with cataclysm when he calculated his ship's position 179 miles off its true course. Hundreds of lives would have been doomed were it not for the fact that the navigator had this time underestimated the distance to port. Where they expected land, they found open ocean. Worse fates have befallen.

To drive his point home, Chappe recalled seven other recent cases of faulty longitudes at sea that resulted in unexpected coastlines committing wholesale slaughter. "Few discoveries, in fact, have more interested mankind," Chappe said. "The discovery of longitudes would preserve for the nation the multitude of citizens who are [instead] buried in the waves."[36]

Contrary to his esteemed colleague, Lalande, newly returned from England, Chappe played no favorites in the longitude war. Indeed, he was eager to contribute to breakthroughs in marine chronometers.

The watchmaker who had returned with Lalande from an only partly successful British spy mission, Ferdinand Berthoud, spent 1764 applying the intelligence he gained to new marine chronometer designs. Berthoud tapped Chappe to join a team that would put the "Montre Marine No. 3" to a sea trial.[37] For once in the bitter race between competing longitude techniques, the chronometer and the astronomer were cooperating.

Chappe and the naval architect Henri-Louis Duhamel du Monceau arrived at the French Atlantic coastal town of Brest—one of the navy's key port cities—in early October 1764. The visitors discovered a shipyard that an uninformed observer might think was in the midst of all-out war. Seventy-four gunships of the line lay in dry dock. Shipwrights, carpenters, and apprentices scurried through the yard refurbishing the gunboats' long hulls and refitting new cannon to their three towering gun decks.[38] Saw pits filled the air with piney scents of fir trees, while the yard's anthill-like activity belied the new peace its nation allegedly enjoyed.

For the first week, Chappe joined two other astronomers at the Naval Observatory in Brest to test the clock's baseline performance on shore— its "ground state," as Chappe termed it. They'd discovered less than Harrison-like performance. One day the clock gained two seconds over a twenty-four-hour period; two days later it gained five seconds.[39] On Sunday, October 14, the team loaded the Montre Marine 3 onto the sixteen-gun corvette *L'Hirondell*. There they set up a glass cabinet housing the chronometer. The timepiece, Chappe reported to the Royal Academy, "is about the size of a coach watch. It . . . can be placed in all parts of the vessel without causing the least embarrassment. Its volume is not [even] a square foot [*sic*]."[40]

Chappe stayed ashore to synchronize the observatory's pendulum clock every day with the exact solar time—while the rest of the team sailed with the the Montre Marine through the coastal waters near Brest. *L'Hirondell* returned after three days. Now, they found, the watch had lost four seconds every twenty-four hours. Next, Chappe joined the crew on a five-night sea voyage. Berthoud's machine this time lost nine seconds every twenty-four hours.[41]

Berthoud's timekeeper performed well, but not stunningly so. Close inspection of the watch's movements revealed that at least its slippage in time was independent of a boat's swaying and jostling. The watch's performance over longer journeys and over greater ranges of temperatures still needed testing.

"Berthoud is certain of the cause that produced these anomalies," Chappe reported to the Royal Academy in Paris in November. Berthoud would be perfecting his design and making his revised Montre Marine 3 available for another sea trial soon, Chappe said. "There is every reason to hope that a second test will leave nothing but elation over the perfection of that clever artist's machines," Chappe added.[42]

All the while, the celestial clock was advancing too. The coming year would see the first plans being laid for 1769 Venus transit expeditions. Here is when Chappe's interest in Berthoud's handiwork stopped. The vicissitude of longitude was a worthy problem of its day—fit for natural philosophers and mechanical craftsmen alike. But the heavens' architecture would only be uncovered once.

And with war now—at least officially—concluded, a true international collaboration like nothing else in human history was getting into gear. Learned men looked now toward a new dawn: June 3, 1769. Astronomers had already forecast that the prime location to view this Venus transit in fact had very little known terra firma from which to view it.

The mysterious and largely uncharted Pacific Ocean would be the new playground for a new kind of player.

Chapter 5

THE BOOK AND THE SHIP

GREENWICH, ENGLAND

September 1765

William Wales and Mary Green were married on Thursday, September 5, 1765, in Greenwich. The bride was the daughter of the late Joshua Green, of the South Yorkshire village of Swinton, and the youngest sister of the astronomer Charles Green. The bridegroom, 30, also hailed from Yorkshire (Warmfield) and worked as a "computer" for the new Astronomer Royal of England, Nevil Maskelyne. Wales regularly contributed mathematical problems and solutions to the "New Queries, Paradoxical-Problems, Rebuses, and Flowers to Be Answered in the Next Year's Diary" section of the annual *Ladies' Diary: or, The Woman's Almanack Containing New Improvements in Arts and Sciences and Many Entertaining Particulars Designed for the Use and Diversion of the Fair-Sex.*

The newlywed Wales and fellow computer John Mapson had a busy season ahead. Since June, they'd been predicting the moon's position in the sky months in advance courtesy of the formulas that Astronomer Royal Maskelyne (appointed to the job in February) had proved in his sea voyages to St. Helena and Barbados. Their job was straightforward, although still difficult. They were preparing the first two editions of a book of tables that would be titled *The Nautical Almanac*. For every day in 1768, they had to forecast where exactly the moon would be at noon and at midnight, Greenwich time. Performing even one such calculation involved hours of mathematical gear grinding, keeping track of coefficients such as the "lunar mean anomaly," "solar mean anomaly," and "the lunar anomaly corrected by the minor equations."[1]

Theirs was a classic cottage industry—albeit one whose product was computations and whose computers were humans. Wales and Mapson— along with Israel Lyons and George Witchell, the lunar forecasters for 1767—worked from their homes, receiving regular mailings from Maskelyne updating them on their assignments.

By July, the Board of Longitude had also hired the Cambridge-based mathematician and astronomer Richard Dunthorne for the job of "Comparer of the Ephemeris and Corrector of the Proofs." Wales, Mapson, Lyons, and Witchell would each individually work out their lunar predictions for the month at hand, either serving as what was then called a human "computer" (calculating each day's lunar position at noon Greenwich time) or "anticomputer" (calculating lunar positions every midnight). Each would post his results to Greenwich, where Maskelyne would examine the pages for obvious errors and then forward the predictions to Cambridge, where Dunthorne considered each prediction in greater detail. If the computer and anticomputer for a given series of days had forecast lunar motions that wobbled or zigzagged (motions the moon does not make), Dunthorne would then sort out where the error originated. And back would go the calculations to the computers and anticomputers to run the numbers again. Once Dunthorne had established the accuracy of the first set of lunar predictions, he then returned the month's framework table to his computers and anticomputers so they could then mathematically interpolate the moon's position at three, six, and nine o'clock, both AM and PM, Greenwich time.

It was painstaking work, but it yielded unprecedentedly precise results.

The *Nautical Almanac* produced longitudes that were an order of magnitude more accurate than its competitors—sometimes down to just 0.017 degrees (1 arc minute) of longitude accuracy for the *Almanac* compared, say, to more than 0.17 degrees (10 or more arc minutes) accuracy for the French *Éphémérides des mouvemens célestes.*[2]

On April 3, 1767, for instance, the *Almanac's* computers predicted the moon's center would be 65 degrees, 39 arc minutes, and 18 arc seconds away from the nearest edge of the sun as of noon Greenwich time. By 6:00 PM Greenwich time, that angular separation had increased—the *Nautical Almanac* predicted—to 68 degrees, 37 arc minutes, and 14 arc seconds. Then, as the moon waxed closer to full (when it would only be visible in the nighttime sky), the *Nautical Almanac* predicted that, for instance, on April 15, 1767, the moon's center would lie 25 degrees, 4 arc minutes, and 34 arc seconds from the star Spica at noon in Greenwich. But by 6:00 PM Greenwich time, the moon separated farther from Spica to 28 degrees, 14 arc minutes, and 26 arc seconds.

The end result was a table that mariners across the world would use to transform the moon into a universal timekeeper. Measuring the angular separation between the moon and sun or reference star gave them, courtesy of the *Nautical Almanac,* Greenwich time. Comparing that to the local (solar) time yielded a quantity—measured in hours, minutes, and seconds—that represented their east-west separation from Greenwich.[3]

Although Maskelyne never received any reward money from the Board of Longitude (while the clock maker John Harrison ultimately won £20,000), the Astronomer Royal's brainchild effectively solved the greatest maritime problem for most ships for at least the remainder of the century. Improvements to Harrison's marine watch would still take decades to render affordable and practicable for the vast fleet of English vessels navigating the planet. But with its January 1767 first edition, Maskelyne's *Nautical Almanac* opened a new and better era of navigation at sea. Any captain or ship's master possessing an £8 Hadley quadrant (or, better still, a £15 brass sextant) could with an additional five shillings buy Maskelyne's two books of tables that saved countless lives and ships from longitude's maw.[4]

Publication of the first *Nautical Almanac*—an annual tradition that began in 1767 and has never missed a year since—marked a historical watershed moment. "What may be termed the pre-scientific age of navigation was brought to a close," historian Eva Taylor wrote. "Landmark or no landmark, the sailor [now] knew precisely where he was—or had the means to know. He did indeed at long last possess the Haven-finding Art."[5]

London
November 1767–February 1768

Human statuettes, animated by clockwork, struck bells over their heads, signifying every fifteen-minute increment on the hour. The west London church whose third-story portico broadcast this clamor, St. Dunstan-in-the-West, vied for pedestrians' attention with nearby Temple Bar. From Fleet Street, whose traffic rushed beneath Temple Bar's archway, a Londoner might admire the Bar's classically inspired lines and curves—or wonder at the two decomposing human heads (examples of the city's rough justice) impaled on spikes above it.[6]

Just up the street from these conspicuous west London landmarks, a lamp hung at the entrance to a darkened cul-de-sac, Crane Court. On November 19, 1767, Nevil Maskelyne walked down the lamp-lit alley and into the small quarters of the Royal Society. The Council of the Royal Society, the scientific body's decision-making board, was now in session. Inside the same Crane Court walls a half century before, Astronomer Royal Edmund Halley had championed Venus transits as "the noblest [sight] astronomy affords."[7] Now newly christened Astronomer Royal Maskelyne endeavored to make his predecessor's vision a reality.

Maskelyne joined the physician and astronomer John Bevis and the instrument makers James Ferguson and James Short to argue, astonishingly, in favor of a glorified crapshoot.

Calculations performed the previous year by Oxford astronomer Thomas Hornsby projected the best locations on the planet to send Venus transit missions. As with 1761, comparing the longest and shortest transit times—combined with a parallel method that compared arctic and tropical Venus transit observations—would yield the most accurate solar parallaxes and, Hornsby said, "consequently the dimensions of the whole solar system."[8]

The longest transit times, Hornsby found, would best be observed above the Arctic Circle in Scandinavian towns such as Tornio and Kittila (in today's Finland) and Wardhus (Vardø, Norway). But the council had learned "that the Swedes or Danes will undertake to make this observation." The Viennese astronomer whom Chappe had met on his way through the Austrian capital in 1760, Father Maximilian Hell, was soon to be outfitted for an arctic expedition to Vardø, at the Danish king's invitation.

A British expedition to Hammerfest, in the northwest Norwegian coastal region, was ultimately mounted too. But the prospects for clouds breaking long enough to observe the sun at this location in early June were bleak. As another British visitor to Hammerfest would chronicle a few decades later, "This island produces nothing: Nature remains in perpetual torpidity—or suffers under the pressure of a perpetual fog."[9] (The expected result came to pass, too. The British team in Hammerfest never saw the Venus transit.)[10]

Although the council resolved to send a backup subarctic mission to the Hudson's Bay colony—and instructed Maskelyne to confirm with his Scandinavian counterparts the veracity of the Swedish or Danish mission—the most pressing question involved the expeditions to observe the shortest transits, those to be carried out near the equator.[11]

Nowhere in the Royal Society Council minutes for November 19 are international politics discussed. But international politics—particularly concerning religion—hovered in the cramped Crane Court room like the "nitrous airs" and "acid airs" the society experimented with in its chemistry demonstrations.

Hornsby had found that the shortest transit could most conveniently be observed from the northern Pacific coast of Spanish-occupied Mexico.[12] To this end, the Royal Society's president James Douglas, Earl of Morton, had lately been corresponding with society member Roger Joseph Boscovich, a polymath scientist based in Milan, to lead the society's Mexican transit voyage. Not only was Boscovich a certifiable genius—having made important contributions to celestial mechanics and the emerging atomic theory of matter—but as a Jesuit, Boscovich carried clout that no English-born Protestant scientist could wield with Catholic Spain.

And had one royal decision not been made, the council might well have handed Boscovich the assignment—transforming his obscurity into the kind of immortal renown that other Venus transit voyagers like Mason and Dixon would soon enjoy. However, tensions within the Catholic world conspired otherwise.

Keeping pace with courts in France and Portugal, in March 1767 the Spanish king Charles III expelled from his country all Society of Jesus members—more than 10,000 people. In exiling his nation's increasingly radical Jesuits, Charles said he only regretted being "too lenient" before.[13]

The Royal Society's surefire plan had just backfired. In May, Douglas wrote to Spain's ambassador assuring him the society knew the suddenly outré Jesuit father would be an "impracticable" choice to lead an English mission through Mexico. Instead, he suggested, "two of our astronomers, delegated by the Royal Society, and each accompanied by an English or foreign servant, [might] go to California to make this important and, in a way, unique observation." In July, the Royal Society heard back from the Spanish Council of the Indies, which was insulted by the society's presumptuousness. The society could no longer rely on Spanish cooperation.

Just twenty-three months remained before an English expedition had to be assembled, outfitted, and shipped to the other side of the planet. The Royal Society still had no firm plans. And so—on this momentous November evening—the society summoned Maskelyne, Bevis, Short, and Ferguson to offer up their best alternatives.

For starters, the Astronomer Royal told the Council of the Royal Society that the French had already picked up the society's dropped lead. Jérôme Lalande, Maskelyne said, had recently made inroads with Spain to send a joint French and Spanish mission to California. So between the Swedes and Danes taking one key transit observation and the French now handling the other, England—home to the very astronomers who first championed Venus transits—might have no substantial role to play in the celebrated measurement.

But, Maskelyne said, some Spanish- and Dutch-discovered islands in the South Seas—the Mendozas (today's Marquesas), Rotterdam (Nomuka, Tonga), or Amsterdam (Tongatapu, Tonga)—could instead serve as a transit mission's Pacific destination. "There is a good harbour in the Mendozas, which is rather to be preferred," Maskelyne meekly added.[14]

Ferguson sputtered on for pages without saying much. ("My opinion is . . . the sun's parallax can be best computed by observations made at those places where the whole transit will be visible.") Bevis and Short made Maskelyne's wild guess of a proposal look concrete by comparison.

"I think it advisable to cross the Tropic [of Capricorn] at about 120° or 130° west of London and then, sailing Westward, to make choice of the first island that offers, provided there be a good harbour and anchorage, fresh water and tractable inhabitants," said Bevis—a seventy-four-year-old doctor whose opinions carried great authority due to his longtime friendship with Edmund Halley.

Short added to Bevis's nebulous counsel that somewhere west of South America and within 25 degrees south of the equator, "a great number of islands are set down in the maps, and any of them will do very well for this purpose."[15]

The Royal Society, in so many words, barely had a clue what to do for its primary Venus transit voyage. And worse, it still needed money.

In a formal petition to King George III in February, members of the society's governing council—including Astronomer Royal Maskelyne, chemistry pioneer Henry Cavendish, and American polymath Benjamin Franklin—projected a 1769 South Seas transit expedition would cost a hefty £4,000 plus the buying and customizing of the ship that would carry the expedition. The council admitted neglect in securing satisfactory results from the 1761 transit expeditions and confessed to a maddeningly uncertain destination.

On the other hand, the council said, English prestige was on the line. Britain was the world leader in astronomy, they said, and "it would cast dishonour upon [the nation] should they neglect to have correct observations made of this important phenomenon." Moreover, "several of the great Powers in Europe, particularly the French, Spaniards, Danes and Swedes are making the proper dispositions for the [transit] observations."

And as the ultimate insurance policy against a case of the royal hohums, the council explained that an accurate set of Venus transit observations would "contribute greatly to the improvement of astronomy, on which Navigation so much depends."[16]

By March, the king had granted the Royal Society its £4,000 plus the Royal Navy's open purse to buy and retrofit a commercial ship for a mission to a yet to be determined island somewhere in the Pacific Ocean.

THE KING'S (SHIP) YARD, DEPTFORD, ENGLAND
March–May 1768

As a top naval architect, Thomas Slade was the mastermind behind some of the most powerful gunboats on the oceans. A student of superior French warship design, Slade stole the Gallic crafts' sleek curves as part of a radical overhaul of the British Navy's battle lines. His seventy-four-gun "third rates" struck a delicate balance between hull-blasting cannon power and steady maneuverability in uncertain seas. A good commander at the helm of one of Slade's ships could play both offense and defense like no others in the world.

His legacy lived long, too. Not only did the Navy continue Slade's designs well past his death—eight of the thirteen ships of the line in Nelson's 1798 victory at the Nile originated from Slade's drafting table—he was also a talented teacher. [17] His students, not least at the Deptford shipyard where Slade learned his craft, became one of the British Navy's secret weapons, working the craft that helped to build the island nation's global empire by century's end.

As the Surveyor of the Navy, Slade pursued a prosaic life. He scrutinized ships' hulls, masts, and yards for maintenance and repair, and—according to the Admiralty's orders handed down on March 23—he sized up a special order for a special mission. The proposed captain of the Royal Society's newly funded South Seas Venus transit voyage, Alexander Dalrymple, joined Slade at the King's Yard in Deptford. The yard, just west of Greenwich, was a hub of Royal Naval activity that had seen history unfold on its very wharfs—from Queen Elizabeth I knighting Sir Francis Drake on his Deptford-docked boat to the Russian czar Peter the Great spending three months there studying shipbuilding.

Dalrymple later wrote in his memoirs that "Alexander Dalrymple [sic] accompanied the Surveyor of the Navy to examine two vessels that were thought fit for the purpose. The one he approved was accordingly purchased."[18]

Dalrymple was a surveyor who'd come to the Royal Society's notice with a recent book he'd written, *An Account of the Discoveries Made in the South Pacifick Ocean Previous to 1764.* In it, he wrote of his admiration for Magellan and Columbus and how "the fond object of his [sic] attention ... was the discovery of a Southern Continent." The mythical

Terra Australis—a hypothesized landmass in the South Seas that might counterweight the vastness of Europe, Asia, and North America—had taunted mapmakers and explorers for more than a century. And Dalrymple was a man possessed, calling the search for the mysterious continent "the great Passion of his life."[19]

A former East India Company clerk who had already ventured through some of the Pacific archipelagos under consideration, Dalrymple had been the Astronomer Royal's favorite pick to lead the transit voyage. There was just one catch. Dalrymple would only participate in the voyage, he informed the society, as its commander.[20]

History does not record how the taciturn, sixty-four-year-old career Navy man received Dalrymple, half Slade's age. But in the pair's review of the two merchant ships—the *Valentine* and the *Earl of Pembroke*—Slade no doubt cast his critical eye as much on the prideful young hotspur as he did on the two barks they were sizing up.

On March 29, Slade and associates reported back to the Admiralty that they'd purchased the *Earl of Pembroke*—"a 'cat-built' bark in burthen 368 tons, 3 years, 9 months old"—for £2,307. Built to haul coal, *Pembroke* was every inch the fatted cow to the majestic warhorses the King's Yard usually sent bounding out to sea. Although she'd ultimately be outfitted with guns, the ship's unwarlike looks would later cause great grief in South American waters. (The term "cat-built" has uncertain origins. One etymology has the first word as an acronym for "coal and timber." Ships of this calling were known for their round bluff bows, their deep waists, and tapered sterns.)

Now Slade and his Deptford apprentices would be transforming a sturdy but dumpy collier into a globetrotting explorer's vessel, ready for anything that might be hiding in the vast South Pacific and beyond.

The three-year-old *Pembroke* was at her prime, having already settled into her joints—but not yet wearing into them or otherwise loosening at the seams. Her bottom and sides, however, had known only cold and choppy North Sea waters. On the other hand, ships anchoring in stiller and warmer tropical ports such as Kingston, Jamaica, had famously fallen prey to marine shipworms, a wood-burrowing insect that, as one traveler said, "cut[s] with great facility through the planks and burrow[s] a considerable way in the substance of them, incrusting the sides of all their holes with a smooth testaceous substance."[21]

Preparing the ship to ward off such infestations was only the first order of duty when on April 7 the Royal Navy's Survey Office recorded, "Ship purchased to be sheathed, filled and fitted for a voyage to the southward. To be called *The Endeavour Bark.*"[22]

Sheathing and filling the outer hull meant slopping on a tar-pitch-sulfur mixture and then fastening fir planks atop the goop. The "fitting," though, was the real challenge. The newly christened *Endeavour*—still little more than a floating box—may have looked ugly. However, the oversize holds designed to maximize coal tonnage could now yield precious living and cargo space that a prettier vessel couldn't accommodate. A collier of *Endeavour's* size might normally carry a crew of sixteen. Slade's men built a whole extra deck, increasing her capacity sixfold to ninety-four men. Still, despite the tight quarters, the spacious officers' area—complete with skylight-illuminated lobby—afforded the kind of elbow room *Endeavour's* botanical and zoological mission required, too. She also sailed with eighteen months' worth of stores: 10,000 kilograms of bread (a dried biscuit called hardtack), 10,000 pieces of salted beef and pork, as well as 5,400 liters each of beer and spirits and 30 tons of fresh water. Despite all the tonnage of equipment and provisions, she nevertheless kept a shallow draft—a standard feature of colliers that allowed them to be beached to unload coal. Such comparatively low displacement would ultimately translate to safe passage through otherwise impassable waters.[23]

Slade's protégés spent the ensuing seventy-two days, minus time lost during a waterfront strike in early May, overhauling the modest bark into a ship that would be enshrined in British naval legend. Incorporation of spare crucial components like anchors and launch boats would prove essential to the mission in due course. One longboat's ultimate destruction at the teeth of tropical marine worms only underscored the importance of *Endeavour*'s proper sheathing and filling. Moreover, even tiny details like backup "gammoning rings" (holding in place the forward-thrusting spar—bowsprit—that in turn helps steady the front mast) and tight covering of the companionway (the quarterdeck's only below-deck access point) might have seemed inconsequential on the Deptford docks. But attention to such details ensured that when the main gammoning did indeed break (off the Brazilian coast) and when gale-force winds submerged the entire quarterdeck (near Cape Horn), the ship's "fittings" proved anything but trivial.[24]

According to one discerning lieutenant who would be circumnavigating the planet on *Endeavour*, the Royal Naval refit of the humble *Pembroke* made the forthcoming voyage "as well provided for . . . as possible, and a better ship for such a service I never could wish for."[25]

The officer was one James Cook. A forty-year-old naval veteran of the Seven Years' War, Cook had first plied the North Seas working in colliers like the *Endeavour*. Now, after having made his wartime reputation as a supremely careful surveyor willing to brave dangerous assignments, Cook had earned the admiration of some of the military's chief officers, such as Navy secretary Sir Philip Stephens. With Stephens and Admiral Hugh Palliser lobbying for him behind the scenes, Cook soon emerged as the Navy's first choice for the mission.[26] But the Royal Society's proposed captain, Dalrymple, would still not relinquish his commanding role.

So Admiral Edward Hawke took a more personal tack. At a Royal Society meeting in early April, Hawke publicly told Dalrymple that "such appointment would be entirely repugnant to the regulations of the Navy." The Navy provided its own helmsmen, Hawke made clear. And those rare occasions when it didn't—such as the near mutiny Edmund Halley had inspired when he captained a 1698 overseas scientific mission—only reinforced the Admiralty's prejudice against civilians commanding military vessels. If Dalrymple was preparing for a standoff, the Navy only saw it as a stand-down. Cook, newly raised to the rank of lieutenant, would captain the *Endeavour*. And that was that. But Dalrymple, the society's records note, "persisted in declining the employment of observer."[27] Lacking a vessel to captain, he would not sail at all.

On May 5, James Campbell—a naval officer and member of the Royal Society Council—suggested his friend Lieutenant Cook could fill dual roles, arguing that Cook was "a proper person to be one of the observers in the observation of the transit of Venus."[28] The society was already well aware of Cook's astronomical qualifications; his careful observations of a solar eclipse in Newfoundland two years before had been the basis for Cook's carefully detailed map of the island. Still, for the upcoming mission, Captain Cook would be second in command astronomically to the new primary observer of the Venus transit mission: Charles Green.

After Maskelyne had risen to Astronomer Royal in 1765, Green—the previous Astronomer Royal's assistant—found work somewhere else. Green had instead pointed his mathematical talents toward a public works venture to divert freshwater from the Coln River west of London toward the city's outer suburbs. No partisan to the counterfactual, Green had told his new employers that surveys of the terrain suggested the scheme wouldn't work. The venture collapsed.[29]

Meantime, in March, Green had married Elizabeth Long in London. And on the strength of his Barbados experience, Green earned an appointment to the post of purser for the Royal Navy's frigate *Aurora*. Despite their falling-out after the Barbados trip, the Astronomer Royal in the end favored talent over grudges. Maskelyne's lobbying assured Green the Royal Society's job of lead astronomer on Captain Cook's Venus transit voyage.[30]

Now with a ship and capable men to captain and lead its primary scientific missions, all the *Endeavour* lacked was a destination. Maskelyne's nebulous best guess was still its best hope.

On May 23, 1768, newspapers began headlining an incredible story. Letters from the newly landed explorer's ship HMS *Dolphin* posted dispatches from a new Pacific paradise. "We have discovered a large, fertile and extremely populous island in the South Seas," the letter read. "From the behaviour of the inhabitants, we had reason to believe [ours] was the first and only ship they had ever seen.... 'Tis impossible to describe the beautiful prospects we beheld in this charming spot; the verdure is as fine as that of England; there is great plenty of live stock, and it abounds with all the choicest productions of the Earth."[31]

Rumors had been spreading that the *Dolphin* crew had brought back Patagonian giants. (The rumors proved to be false.) Stories of first encounters with the natives of the new island—Tahiti—did, however, withstand verification. "The first day they came along side [the *Dolphin*] with a number of canoes, in order to take possession of her. There were two divisions, one filled with men, and the other with women—these last endeavoured to engage the attention of our sailors by exposing their beauties to their view," a correspondent wrote in the *London Magazine*. After describing an initial failed attempt at attacking the *Dolphin*, one that left the ship's captain Samuel Wallis no recourse but to open fire with her big guns, the peoples of this island "immediately showed the greatest desire of being at peace with us."

"The natives," the correspondent continued, "are in general taller and stouter made than our people and are mostly of a copper colour with black hair.... It does not appear that they know the use of any one metal whatever. When the grapeshot came among them, they dived after it and brought up the pieces of lead."[32]

PLYMOUTH
May–August 1768

On May 27 the *Endeavour*'s masthead first carried a white and red Royal Naval pendant, signifying an officer of His Majesty's fleet now commanded the ship. Lieutenant—the Royal Society more loosely termed him "Captain"—James Cook now was at the helm. Since taking charge of the ship, Cook had supervised *Endeavour*'s final renovations, had received ordnance downstream on the Thames (along a stretch of river called Gallions Reach), and had sailed to Plymouth for her final loading and staffing before setting out to sea.

The onboard scientific kit that Cook and Green were preparing constituted some of the finest portable astronomical instruments in the world. Royal Society member James Short—who'd essentially shrugged his shoulders when asked where to send the transit voyage—made the most sought-after telescopes in England. He furnished two for *Endeavour*. The Navy also provided Cook with a simpler scope with which he'd already become familiar on previous missions, while a British Museum scientist on the voyage would be bringing his own thirty-six-inch telescope to rival the two-footers Short had built. For measuring angles between objects in the sky, the society outfitted *Endeavour* with the best designs of dimensions both compact—brass sextant by Jesse Ramsden—and large—one-foot astronomical quadrant by John Bird.[33] The Astronomer Royal's staff had been working overtime to supply *Endeavour* with *Nautical Almanacs* for both 1768 and 1769—enabling the quadrants and sextants to serve as precision longitude finders for at least the next year and a half. And for tracking the ship's heading, the Navy installed magnetic compasses from the shop of Gowin Knight, the nation's most celebrated geomagnetist.[34] Similarly top-rated and redundant clocks, thermometers, micrometers, and stands rounded out a gear manifest that carried the pride of an increasingly technologically sophisticated nation. The Royal Society was also sending the *Nautical Almanac* computer William Wales and assistant Joseph Dymond to Hudson's Bay to observe the Venus transit. (This despite the fact that Wales had informed the society that he "prefer[red] a voyage to a warm climate.") No less impressive than *Endeavour*'s scientific instruments were Wales's James Short–designed telescopes as well as a pair of quadrants, three clocks, a barometer, and a thermometer.[35]

Royal Society fellows weren't the only ones pushing the technological limits of the day. The 1761 transit had captured the public imagination, and the 1769 sequel was becoming the hot new avocation of scientific enthusiasts in both Old World and New. Another sixty British observers from Martinique to Philadelphia would also be reporting their transit measurements back to London and Paris to be weighed alongside the big-budget expeditions. All told, these secondary observers—with kindred floods of amateurs in Sweden, France, Russia, Germany, and Spain—created a booming new marketplace for instruments of exactitude. Some non-transit-related business, in fact, ground to a halt. The Scots explorer James Bruce, traveling at the time through Egypt to trace the source of the Nile, discovered that he couldn't replace an astronomical quadrant he'd lost in a shipwreck because, he wrote, "all the excellent instrument makers in Europe [were] employed by the astronomers of different nations then much engaged about the transit of Venus." Bruce instead borrowed an old quadrant from the court of King Louis XV of France. [36]

Launching in a time of peace—and with a globetrotting mission like nothing before it—*Endeavour* drew on a talent pool equally as impressive as the gadgets being stowed in her holds. Senior lieutenants, medical staff, and midshipmen all brought with them years of expertise earned on the high seas. Four officers on Cook's quarterdeck had logged firsthand experience at *Endeavour*'s destination. Cook's third in command, John Gore, had circumnavigated the planet twice—once on Wallis's "discovery" mission to Tahiti (or as they dubbed it at the time, St. George's Island) and once before under Commodore John Byron. Three of Cook's surveyors and navigators had also known St. George's Island—Robert Molyneux, Richard Pickersgill, and Francis Wilkinson. Molyneux, whom Cook described as "a young man of good parts but ... given ... to extravagency and intemperance," was just one of the libertines who had found the South Pacific a welcome outlet for their vices.[37]

Legends of St. George's Island, having scarcely a month to warp into full-blown myths, had already pricked English ears. Molyneux, Wilkinson, Pickersgill, and Gore were all well qualified to launch whisper campaigns among *Endeavour's* crew of the tropical—and sexual—paradise that awaited.

The musket balls HMS *Dolphin* had loosed on St. George's natives created a sudden aboriginal marketplace for all things iron. And St. Georgian women discovered a valued commodity they could trade with lonely English sailors. Before casting off, most of the *Endeavour* knew of the *Dolphin's* crew's encounters—tales that would later appear in print. "The [*Dolphin's*] carpenter came and told me every cleat in the ship was drawn and all the nails carried off," *Dolphin's* master George Robertson recorded in his ship's journal. "I immediately . . . called all hands and let them know that no man in the ship should have liberty to go ashore until they informed me who drawed the nails and cleats— and let me know what use they made of them. But not one would acknowledge that they knowed anything about drawing the nails and cleats, but all said they knowed what use they went to."[38]

Sex in exchange for nails. What Robertson soon euphemistically referred to as the "old trade" had effectively ripped the *Dolphin* to pieces. Only strictly enforced crackdowns and abstinence regimens saved *Dolphin* from falling apart in the harbor. But venereal disease had already begun to spread among the ship's hands like a strain of flu.

Lieutenant Cook ordered his ship's carpenter to stow an extra barrel of nails on *Endeavour's* voyage, just in case.

Prowess of more than one kind would soon be coloring the *Endeavour's* travels. A rich explorer named Joseph Banks would also be embarking in *Endeavour* with a complement of seven associates who shared Banks's enthusiasm for botanical discoveries and perilous adventure. Cautious family members and friends urged Banks to reconsider the voyage and instead undertake a grand tour of Europe. "Every blockhead does that," Banks replied to one such appeal. "My Grand Tour shall be one round the whole globe."[39] It was an expensive Grand Tour too. Banks would ultimately funnel in £10,000 of his own money to *Endeavour's* round-the-world voyage.

By June, London newspapers had seized on the romantic—and perhaps perilous—prospects of the *Endeavour*'s mission. "Several astronomers are going out in her to observe the transit of Venus over the Sun," *The St. James's Chronicle* reported, "and some gentlemen of fortune who are students of botany are likewise going in her upon a tour of pleasure. Thus we see that a voyage round the world, or to the South Sea, which a few years ago was looked upon as a forlorn hope—and the very mention of which was enough to frighten our stoutest seamen—is now found from experience to be no more dreaded than a common voyage to the East Indies."[40]

The Admiralty also handed Cook a sealed packet of secret orders that he was to open only upon completion of the Venus transit portion of the trip. Banks suspected that the orders involved heading southward from St. George's Island to scour the South Pacific for the mythical missing southern continent.

Half of the planet needed to be navigated, though, before anyone could worry or argue about any great lost landmass of the Southern Hemisphere. The Atlantic Ocean, for one, held mysteries aplenty still.

Setting out from Plymouth on Friday, August 26, 1768, Cook recorded a matter-of-fact entry in his captain's log. "At 2 p.m. got under sail and put to sea having on board 94 persons including officers, seamen, gentlemen and their servants," Cook wrote. "Near 18 months provisions."[41]

Banks summed up *Endeavour*'s mood of excitement and foreboding when he wrote in his journal fifteen days later: "Today for the first time we dined [off the coast of] Africa and took our leave of Europe for heaven alone knows how long, perhaps for ever. That thought demands a sigh as a tribute due to the memory of friends left behind. And they have it. But two cannot be spared. T'would give more pain to the sigher than pleasure to have sigh'd for. 'Tis enough that they are remembered. They would not wish to be too much thought of by one so long to be separated from them and left alone to the mercy of winds and waves."[42]

So departed a bark that would become one of the most legendary sailing ships of its time. However, *Endeavour's* primary mission remains far less celebrated than the captain who stood at its helm. Unjustly so, too. For the legendary Captain Cook would not be the legendary Captain Cook without the Venus transit that first sent him sailing across the planet. Venture now, one might paraphrase Cook's orders, to the ends of the earth to find the nearest faraway star.

Chapter 6

VOYAGE EN CALIFORNIE

Paris
January 1768

Jean-Baptiste Chappe d'Auteroche had finally, after five years' preparation, completed his memoir of the Siberian expedition. Lesser explorers might have confined their remarks to the adventurous journey and the Venus transit observations—perhaps tossing in a few lyrical odes to the higher calling the transit affords for knowing mankind's place in the cosmos.

Chappe, on the other hand, produced a manuscript, "Voyage en Sibérie fait par Ordre du Roi en 1761," in which Venus's rare conjunction with its parent star was just a launching point. Everything under, and including, the sun was his bailiwick. And wherever great mysteries or controversies lie—or could be stirred up—Chappe directed his pen.

"Nature has beauties even in her horrors," Chappe would later write, in a lyrical passage that encapsulates his whole philosophy.

> Nay, it is there perhaps that she is most admirable and sublime. The calmness of a fine day is in some measure less interesting than the moments of distress, when the waves, lifted up by the

winds, seem confounded with the sky. Deep gulfs are opening at every moment. At this instant man shudders at the sight of a danger that appears inevitable. But anon, when he sees the calm succeed the tempest, his admiration turns upon himself, upon the vessel, upon the pilot, who are come off conquerors over the most formidable elements. A secret pride rises in his mind. He says within himself, "If man, as an individual, is but a speck, an atom in this vast universe, he is by his daring spirit worthy to embrace its whole extent, and to penetrate into the wonders it contains."[1]

Chappe's eloquence, as expressed above in a poetic description of the Western scientific worldview, reveals how undeserved is his obscurity today. Though he didn't live as long as his American counterpart, Benjamin Franklin, Chappe's star burned brighter and shone more tremendous insight and character than any other Venus transit astronomer of the 1760s. May he someday find welcome among the Enlightenment's luminaries who are indeed his peers.

In addition to the many controversial reforms Chappe prescribed for Russian society, thrown into the finished manuscript, the natural philosopher also pondered an emerging mystery with theological implications: fossils. Biblical scholars who had carried the Old Testament's chain of begats back to Adam and Eve calculated that at exactly 9:00 AM on October 26, 4004 BC, God created the heavens and the earth. Many natural philosophers accepted the Bible as prima facie evidence of the origin of man, beasts, the universe, and their home planet. Even such preeminent geniuses as Isaac Newton and Leonhard Euler accepted the literal truth of Bible stories—and endeavored to use the tools of science to better understand the workings of Scripture. Others such as Chappe, however, proved less eager to seek out scientific verification of biblical legends.

How a man read a fossil, in other words, revealed something about how he read God. For instance, in Chappe's home city the famous author Voltaire had been wrestling with various accounts of marine fossils newly discovered at high elevations. These might be seen as proof of Noah's flood. But Voltaire, whose deist theology questioned biblical myths, didn't believe it. On the other hand, the natural history cabinet at St. Petersburg's Academy of Sciences—where Chappe had lectured about the Venus transit in 1762—kept fossilized remains of a rhinoceros found on the tundra that local experts interpreted as "convincing proof . . . [of] a most violent and rapid inundation which formerly bore such carcasses toward these frozen climes."[2]

During his Siberian odyssey, Chappe had discovered pieces of a tusk which, he wrote, "must have belonged to an elephant of the largest kind." At a recent gathering at the estate of the learned courtier François Le Tellier, Marquis de Courtanvaux, Chappe had met his American doppelgänger, Benjamin Franklin—each man becoming known in his home country as an outspoken controversialist and intellectual omnivore. Franklin too had recently come into possession of elephant remains, in his case from the Ohio River valley. Neither saw the out-of-place pachyderms as roundabout evidence for a biblical flood. Instead, Chappe and Franklin simply wanted to follow the evidence to whatever conclusion it might lead, heretical or no.

In January 1768, Franklin sent Chappe an elephant tooth that perplexed the American. Franklin said in his accompanying letter to Chappe that some colleagues in England suggested the tooth could only have been useful to a carnivorous animal, and thus could not belong to an elephant. But, Franklin wrote, "those knobs [on the tooth sample] might be as useful to grind the small branches of trees as to chaw flesh. However, I should be glad to have your opinion and to know from you whether any of the kind have been found in Siberia."[3]

At the same time, even as Chappe's three-volume account of his Siberian adventures began making a name for the explorer in his home city, the Royal Academy of Sciences began preparing to take him out of it again. This time, though, a frigid destination was not on the program. Chappe's fellow cleric-astronomer Alexandre Guy Pingré had, at meetings of the Academy, been a vocal proponent for a South Seas destination this time. As with British preparations for the 1769 transit, French scientists sought sites that offered the shortest and longest transit times—and thus, by comparing the two, would ultimately yield the best solar distance ("solar parallax"). The longest times would be available, Pingré said, somewhere in northernmost Scandinavia. And, Pingré added, "the Swedes and the Danes will penetrate into the most favorable places in Lapland."[4] The Nordic kingdoms, in other words, would take care of it.

So, Pingré asked, where should the academy send an expedition to provide the ideal complement to the Swedes' and Danes' data? "Easter Island," Pingré said, "is best placed of all to effect this parallax."[5] The astronomer then went on for pages poring through recent accounts of other South Seas adventures for other possible backup observing stations.

Members of the academy had been speaking with Chappe about venturing to Easter Island—or some other remote Pacific outpost—not just to collect Venus transit data. They spoke with Chappe about undertaking the same mission to explore and chart these strange new worlds and to put the latest French marine chronometers to a sea trial.[6]

LE HAVRE, FRANCE
September 1768

Chappe's ship, *Le Nouveau Mercure*, could have been any of the dozens of cruisers that passed through this French port city on any given month. Her provisions—contrasted to the small town stuffed into the HMS *Endeavour*—represented no great stowage. A ship setting sail for Easter Island with this cargo, like ascending the Alps with a change of shirt and a couple apples, would barely have made it to the eastern shores of South America.

But plans had changed. Chappe and his three fellow voyagers were still carrying kit enough to construct a world-class scientific lab wherever they landed. Beyond that, the voyage launching from Le Havre's quayside was nothing remarkable. Instead, passage to their Venus transit observing site—on the southern tip of the Baja peninsula—was now Spain's duty. *Le Nouveau Mercure* needed only to make it as far as Cadiz, a Spanish coastal town near Gibraltar where much of the Iberian nation's fleet laid anchor.

Spain had effectively taken over the expedition because it had colonized the prime transit-observing destinations where the French wanted to travel. Yet the king of Spain, Carlos III, wasn't interested in underwriting an expensive voyage to a remote Pacific island. Instead, the proposal the British Royal Society had already broached—observing from the Baja peninsula on Mexico's Pacific coast—soon became the king's preferred plan. Still, France had the best astronomers. "It was thus decided that M. Chappe would go to California," Chappe's colleague Jean-Dominique de Cassini later wrote. "And he would situate himself the closest he could to the southern tip of the peninsula of Cabo San Lucas—to have the shortest [transit] time possible." Chappe would not be observing alone, however. King Carlos also instructed his scientific authorities in Cadiz to nominate two Spanish astronomers to join Chappe on the voyage across Mexico.[7]

Cadiz, Spain
December 1768

Each idle week brought a new insult. Chappe had departed from Le Havre on September 27, delayed enough as it was. Moreover, the journey to Cadiz had consumed an additional week and a half of unexpected travel time. "A hard gale that we met with north of Cape Finisterre left the sea very tempestuous for near a week after," Chappe recorded in his journal. Yet still the lion's share of land and water yet remained to be crossed—from the Atlantic Ocean to the thousand-mile overland journey from Veracruz via Mexico City to the Pacific port town of San Blas and then across the Gulf of California to a yet to be determined site on the southern Baja peninsula.

However, upon arriving at Cadiz on October 17, Chappe and his three traveling companions discovered that Spanish bureaucracy would be delaying them even longer than the waves and wind off Finisterre had.

The orders from Madrid had only specified Chappe as France's Venus transit voyager. The group's sixteen-year-old artist, Alexandre-Jean Noël—tasked to record the trip's zoological, geographical, and botanical finds—was not allowed to board. The same rebuff of omission withheld Chappe's astronomical assistant, Jean Pauly, and the technician, "M. Dubois," trained to repair all the clocks and instruments. Worse still, only one such instrument was allowed on the voyage. Chappe had five telescopes—two conventional models, a "transit instrument" for observing celestial objects at their highest altitudes, a "mégamètre" for simplifying lunar longitude measurements, and one in the vain hope of observing the moons of Jupiter at sea—thereby, in such a pipe dream, solving the longitude problem. Chappe's group also carried two quadrants, the latest version of Ferdinand Berthoud's experimental marine chronometer, as well as a precision compass, barometer, thermometer, and a specially designed prototype instrument for measuring water density that, as another long shot, was hoped might yield more useful proxies for longitude at sea.[8]

So, with the ocean air taunting his nostrils every day, Chappe spent the final two months of 1768 stuck in this Spanish port city, lobbying for and then awaiting the paperwork that would let him and his team proceed as originally planned. Initial hopes of sailing on a Spanish galleon soon ratcheted down to practically any seaworthy craft. As if urging the travelers to keep their eyes on the horizon, Cadiz offered nothing to see onshore—its unfinished cathedral still hadn't cleared thirty feet—with the only walking path in town providing unobstructed views of, as one of Chappe's colleagues wrote, "the ships continually going in and out."[9]

By mid-December, Chappe had had enough. "When I calculated the time it would take to reach ... California," he wrote, "I foresaw it was morally [*sic*] impossible we should get there in time for our observation, if we were retarded ever so little longer." The Frenchman notified his ambassador at Madrid that his team now needed to take "the first ship, no matter which" that was bound for Vera Cruz.

Finally in mid-December, Spanish officials caved. A tiny craft, with a crew of only twelve, would ferry Chappe, his assistants and gear, and his Spanish crew and co-observers across the Atlantic. "The frailty of the vessel I was going to venture in, and on which account some people endeavoured to intimidate me, was in my eyes one merit the more," Chappe recorded in his journal. "Judging of her swiftness by her lightness, I preferred her to the finest ship of the line."[10]

Vera Cruz, New Spain (Veracruz, Mexico)
March 1769

Chappe's seventy-seven-day transatlantic voyage was only remarkable for its predictable difficulties. His small brigantine craft, he wrote, encountered "every kind of weather, calms, storms, winds, sometimes fair, sometimes contrary; such is in few words the history of most voyages; and as to ours, we may add, a continual tossing of our little nutshell, which was so very light as to be the sport of the smallest wave."

The tedium of constant motion set Chappe to thinking of a line from the Roman poet Horace: "That man must have had a heart cased in oak and three-fold brass who first committed himself in a frail bark on the raging sea."

This quote from antiquity, he wrote, "is what I repeated a thousand times on our voyage, thinking of Christopher Colomb, [Juan de] Gryalva, and all those first intrepid mariners, who in quest of a new world, upon a mere surmise of its existence, suggested by their own genius, dared to undertake near-three-hundred years ago those very voyages which at this day we still account dangerous, though assisted with a thousand helps that were wanting in the days of those great men."[11]

For all his eloquence, Chappe was not a poet. He mostly busied himself during the passage in taking barometer and thermometer readings and in measuring seawater density throughout the trip.[12] Chappe also tested the prototype telescope he'd been given that was designed to discover longitude at sea via timing of eclipses of the moons of Jupiter. And like Charles Green and Nevil Maskelyne's disastrous results five years before with the English "marine chair"—a telescope engineered with the same goal in mind—Chappe proved the utter folly of such an enterprise.

Last, Chappe tried taking lunar longitudes—aided by the "mégamètre" telescope that allegedly eased the measurement of angles between moon and reference star. But measurement wasn't the hard part of lunar longitudes. Calculations were. And without a *Nautical Almanac* to fall back on—an almanac that had already computed most of the difficult calculations in advance—Chappe convinced himself that only marine chronometers could ever practically yield longitudes at sea. "The tedious calculations which this method requires, with the accuracy and attention requisite in the observation itself, make it doubtful to me whether it will ever be fit for the use of trading vessels," Chappe wrote.[13]

On March 6, Chappe and his men—having watched the choppy ocean swallow up most of their bobbly ship's store of food and drink—set anchor five miles outside the harbor at Vera Cruz. Access to this entrepôt of New Spain was famously difficult from the water. Rough winds from the north this time of year whipped up a fury of complications compounded with reefs and sandbars that challenged even experienced pilots.[14]

Typically a ship approaching Vera Cruz harbor would fire her guns twice, indicating she needed a pilot to guide her through the treacherous waters. Chappe's captain, instead, ran up the French flag as his request for help. "This," Chappe dryly observed, "was the ready way to get no assistance."[15]

A day of thwarted attempts to negotiate jagged coral complexes like the Reef of Galleguilla and the Reef of Isla Verda only whittled down the fast-depleting provisions.

Then, on March 8, a cannon shot from Vera Cruz's fort plus spoken orders delivered by a harbor-cruising sloop suggested they'd better find a place to anchor immediately. The north winds were picking up. And the channel that Chappe's brigantine ultimately had to settle for was narrow enough to risk foundering the whole mission on nearby rocks.

A joint French-Spanish expedition sinking in Vera Cruz harbor might not have risked any actual lives. But the harbor incident was still a pivotal moment. Chappe helmed the Bourbon dynasty's most ambitious plan to secure the other crucial number—in addition to transit observations the Swedes and Danes were now being counted on to provide—that could cast humankind into a universe of known depth. England was staking its lot on a long-shot mission into largely uncharted South Seas waters. Who knew if the English vessel would ever be heard from again?

Landing at Vera Cruz and following familiar trade routes through New Spain, on the other hand, was nothing so outlandish. Attempting a surefire mission like Chappe's and failing would have proved both an embarrassment and a disgrace—an insurance policy that never insured.

Little surprise, then, that when Vera Cruz's governor learned that the nutshell bobbing out in the harbor carried the king of Spain's imprimatur, he ordered a safer berth for the boat. The mission's two Spanish co-observers, Salvador de Medina and Vicente de Doz, first offloaded onto a harbor vessel to port them to shore. Two hours later, another craft came for Chappe and his assistant Pauly.

The sky loomed. The waves slapping against the boat's sides and spraying its passengers were only increasing in size. As rain began pelting the travelers, Chappe's craft landed and offloaded him into the dank little city founded by the legendary conquistador Hernán Cortés. Little more than a decaying way station for Mexican loot, Vera Cruz was hardly the most cosmopolitan of cities. "The town is not very considerable either in point of size or the magnificence of its buildings," a contemporary visitor to Vera Cruz observed. "For on the one side being exposed to vast clouds of dry sand and on the other to the exhalations of very rank bogs and marshes, it is so very unwholesome that scarce any Spaniard of note resides there constantly."[16]

Still, it was land. And considering the sky's outpourings, land of any kind was becoming more and more welcome by the minute. Chappe may not have been fluent in Spanish, but he didn't need proficiency to understand the import of the chatter in town: "Huracán!"

Now cut off from his assistants and some of the finest scientific instruments available anywhere in New Spain, Chappe watched helplessly as the storm tides and pounding winds battered his brigantine. A wise hand onboard the craft maneuvered it into the lee of the nearby island fortress, San Juan de Ulua. The tempestuous waters mercilessly tossed the little nutshell and would surely have crushed it had the pilot not sought shelter. But shadowed by the mighty stone castle, the passengers and precious cargo were spared to continue their journey, now venturing inland.

Mexico (Mexico City)
March 26–29, 1769

On his eight-day cross-country trek to the capital city of New Spain, Chappe had seen enough of Spanish rule to understand how it worked—and how its boot resembled the Russian model the Siberians lived under.

Despite the colossal plunder conquistadores were extracting from New Spain, Chappe attended a mountainside fair near the high-altitude hamlet of Xalapa that conspicuously lacked shiny metals. "The Mexicans give in exchange cochineal [dye] and money, for as to gold or silver bullion, no body is allowed to have any," Chappe recorded. "A breach of the regulations respecting the mines is the greatest crime that can be committed. A false coiner is hanged; a murderer is only imprisoned or banished."[17]

The entourage consisted of two sedans—one containing Chappe and Pauly, the other containing their Spanish counterparts Doz and Medina—preceded by a mule train that the rest of the crew either rode on or rode herd over. As Easter weekend approached, the train reached the outskirts of Mexico City. The roads beyond Xalapa led toward a peak that provided, Chappe said, "a most singular prospect: We stood so high that the clouds were our horizon." The aromatic spring air and lush scenery on the way down heightened the contrast to the scenes of human misery before their eyes. The Spanish colonial forces had put what remained of the original population to work extracting mineral resources. "The ill treatment these poor Indians receive from their masters contributes as much as sickness to destroy the race," Chappe wrote. "And the mines where they make them work yearly prove fatal to an infinite number of these poor wretches. . . . The province of Mexico is now but a vast desert compared to what it was in the time of Montezuma."[18]

At noon on Easter Sunday, March 26, Chappe and his mule train arrived at the capital of New Spain. The colony's viceroy had ordered his men to forgo the routine inspection of the visitors' baggage. The distinguished guests were led to a spacious converted cathedral that would be their lodgings—a cathedral that Chappe learned had been seized from the Jesuits when Spain expelled the order from its territories two years before.

The man who carried out the expulsion, Viceroy Carlos Francisco de Croix, was a former army general who made no pretenses about his preferences. For whatever unspecified reasons, de Croix took a liking to Chappe and company. "I am at a loss for words to express [his] friendship and politeness," Chappe wrote in his journal, adding that de Croix "left nothing undone to procure us whatever we wished for. . . . We had no table but his own for the four days we continued in the town, and he was so obliging as to send a cook to dress victuals for our attendants after the French fashion."[19]

Just five hundred yards southwest of their accommodations, Chappe took a walking tour of the prime jewel of a city overflowing with rich mining families and their pillage. Mexico's Metropolitan Cathedral (Catedral Metropolitana) shone with the glint of an encrusted crown. Silver rails surrounded the main altar, which barely contained an oversize lamp of solid silver enriched with ornaments of pure gold. Gold-fringed velvet hangings streamed down from the inside pillars. But the ground beneath the cathedral, the silty bottom of a drained lake, was starting to give way. "The outside of the cathedral of Mexico is unfinished—and likely to continue so," Chappe recorded. "They are afraid of increasing the weight of the building, which already begins to sink."

While de Croix was known for his comparatively even-keeled administration, all was relative in New Spain. The Inquisition still burned penitents in the city's nearby site of religious torture, the notorious *quemadero*. Seventeen heretics had paid the ultimate price in 1768, another four the year previous. "This is the place where they burn the Jews and other unhappy victims of the awful tribunal of Inquisition," Chappe wrote in disgust. "This quemadero is an enclosure between four walls and filled with ovens, into which are thrown over the walls the poor wretches who are condemned to be burned alive—condemned by judges professing a religion whose first precept is charity."[20]

In a city that rewarded ostentatious cruelty, Chappe had found a few kindred spirits during his four days in town. Most productive scientifically was Chappe's meeting with a similarly omnivorous mind—the Mexican polymath priest, scientist, historian, cartographer, and journalist José Antonio de Alzate y Ramírez. Although Chappe did not record his personal interactions with Alzate, the Mexican philosophe drew up a detailed map of the New Spanish dominions Chappe's team would be exploring. Alzate also shipped a chest full of Mexican natural history specimens for further study to the Academy of Sciences in Paris—specimens of local trees, fruits, leaves, fishes, butterflies, shells, crystals, rocks, ores, seeds, and flowers that Alzate and his French counterpart probably spent some time discussing during the visit.[21]

Chappe had also discovered a native Frenchman living in the city who spoke both Spanish and indigenous Mexican languages. "I took him for my interpreter," Chappe wrote, "as I thought he would be very serviceable to me for the remainder of our journey, and especially in California."

Unlike the heavily traveled roads connecting the capital city to its principal Atlantic port, the roads ahead carried fewer travelers and more hazards. "The viceroy thought proper to give us a guard of three soldiers to defend us against the robbers who infest those parts," Chappe noted. "Troops of fierce and unconquered Indians, called by the Spaniards, Indios bravos, attack travelers when they find themselves strongest, murder them—or at least, after stripping them and tying them to the neighboring trees, they carry off their mules and baggage."

The Mexican soldiers now joining the expedition as it forged westward on March 30 said that any Indios bravos the group encountered would not be handled peaceably. "Our guides told us . . . these banditti are easily known by a handkerchief [that hides their face]," Chappe wrote. "When a traveler sees an Indian thus masked, the safest way is to be beforehand with him, and to kill him if possible."[22]

Outside Querétaro (Santiago de Querétaro, Mexico)
April 1769

Knowing the roads ahead would be rugged, Chappe chose horseback for his trek to San Blas—the Pacific port town that would be their launching point to the Baja peninsula. Doz and Medina opted instead for the same "litter" carriage on which they rode into the capital city. Chappe knew his equestrian mount didn't make the journey any easier for him. But, he wrote, "I escaped a thousand mischances that befell our Spanish officers."

Images of danger and abject poverty dogged the voyagers on their 650-mile passage west of New Spain's capital. "The farther you go from Mexico [City], the fewer habitations you meet with, and the road is often very rough, dangerous and full of precipices," Chappe recorded. "In most places where we stopped, we hardly found bread, and every thing in that part of the country wears the face of the most pinching penury."[23]

Some 140 miles west of Mexico City, Chappe paid witness to a high-altitude phenomenon that he'd also recorded during his Siberian voyages: lightning strikes that begin at the ground and then rise up to meet the storm cloud. A rare phenomenon now called "positive lightning," these unusual electrical discharges were in Chappe's day considered a possible cause of earthquakes.[24]

Chappe's characteristically childlike sense of awe at this natural wonder set his musings in marked contrast to the scientific papers of many contemporaries. The voyaging French astronomer displays throughout his volumes no loss of objectivity when his imagination is truly captured. At the same time, vivid language seems readily at hand for him to describe in subjective terms the visceral excitement he's clearly feeling.

"I observed to the south a great black cloud at a moderate height above the horizon; the whole hemisphere about us had a fiery aspect," Chappe recorded on April 3. "All the while it remained in this state, frequent and smart flashes of lightning appeared in three places of the cloud over these columns. . . . Soon after, the cloud came lower down, and then it was that we saw incessant lightnings rise like so many sky rockets, and flashing at the top of the cloud."[25]

The Gulf of California
April–May 1769

By now, the stories of previous transit voyages—and their lessons learned—had traveled around the world. In 1760 a fellow member of the French Academy of Sciences, Guillaume-Joseph-Hyacinthe-Jean-Baptiste Gentil de la Galaisière ("Le Gentil"), had set sail from France on a Venus transit expedition that took him 15,000 miles around Africa—evading chase, along the way, from a British man-o-war—and to Pondicherry, a French colonial outpost in India. But just before Le Gentil's arrival, the British had taken Pondicherry. So without a place to land, Le Gentil spent June 6, 1761, floating somewhere in the Indian Ocean—unable to set up a telescope onboard his swaying ship to observe Venus traversing the face of the sun. Le Gentil had ventured halfway across the globe, in other words, for nothing.

With less than a month to go before the 1769 Venus transit, Chappe lay adrift in his own watery purgatory, recording in his journal increasingly anxious entries about the "calms and currents" on the Gulf of California. He had, in fact, only received what he had been told to expect.

On April 15, upon arriving in San Blas, Chappe grew nervous that he might not make his Baja destination in time. The Californian gulf was well known for its quiet calm. The captain of Chappe's new packet boat, *La Concepcion,* said his last crossing of the deceptively narrow three-hundred-mile passage took twenty-one days. And that was at a better time of the year for favorable winds. So Chappe and his fellow Spanish astronomers weighed a last-minute alternative—setting up their observatory in San Blas and observing the Venus transit from there.

But local officials in San Blas informed the travelers that the wet season was fast approaching. Steady rains usually arrived in late May and remained hunkered over this New Spanish Pacific port town for the ensuing month. The nearby Tres Marías islands, seventy-five miles offshore, offered no better prospects for clear skies on June 3 either.

So the voyagers decided to risk it. Setting sail on April 19, the packet boat spent the next fifteen days trying in vain to fight contrary winds and currents. From May 4 to May 9, a breeze at last carried them north-ward along the shoreline to the latitude of their Baja destination. But they still had another two hundred miles of longitude yet to go. They still needed to actually cross the gulf.

With just twenty-five days remaining before the century's final Venus transit—during which time Chappe and his crew would also require at least a week to set up their instruments and observatory—eyes looked westward. And hearts sank. Even if *La Concepcion* had picked up her pace to the rate the captain had reported on her previous crossing, they'd still not make it in time.

The ship's captain, no doubt in a gesture intended to placate the founder of the recently expelled Jesuits, set up a shrine on *La Concepcion* to St. Francis Xavier. The captain, Chappe wrote, "laid [an offering] upon the binacle, beseeching him to send us a fair wind. The devout pilot's remedy did not presently take effect, for the following days we had a succession of calms and contrary winds."

The boat continued to sail farther north, beyond the latitude of its destination, hoping to find a more generous current to cross the gulf. "From this time, it was my fixed resolution to land at the first place we could reach in California," Chappe recorded. "I little cared whether it was inhabited or desert, so [long] as I could but make my observation."

By nightfall on May 18, some "favorable gales," in Chappe's words, had brought *La Concepcion*—almost miraculously—to within twenty miles of Baja peninsular land. The captain surmised they were ap-proaching San José del Cabo, a small town near the peninsula's southern tip. The Spanish officers knew, however, that San José del Cabo would be a difficult landing. Heading toward this patch of Baja, they argued, risked wrecking the whole ship. "I was strenuous for landing at the near-est place," Chappe recalled. "But as I was singular in my opinion the whole day was spent in altercations. . . . I was confident that his Catholic Majesty had rather lose a poor pitiful vessel than the fruits of so important an expedition as ours."

Doz, Medina, and the captain argued for traveling a little farther down the coast, to more accommodating ports in the bay of San Barnabé. Chappe would have none of it. Time was running out, and no options looked better than the coastline that now lay before them. The ship's master, familiar with San José del Cabo, said that although the landing might be rough, he also knew a Franciscan mission nearby that could serve the expedition's purpose well.

So on Friday, May 19—with just over a fortnight before the Venus transit—*La Concepcíon* dropped anchor less than two miles from the river that led inland toward their ultimate destination. As if on cue, the wind whipped up a new microstorm, one that set tempers blazing again. But it died down just as quickly, before any heated verbal exchanges could be logged.

Pauly and the expedition's young artist, Alexandre-Jean Noël, climbed aboard a longboat and hauled most of the equipment ashore. The sight of such a pathetic craft ferrying such essential gear must have raised the nerves to the kind of heights that a man like Chappe, under other circumstances, might have wanted to study. Chappe could only watch helplessly as the longboat capsized again and again in the rough surf.

Pauly returned to *La Concepcíon* alone, informing his boss that through some brave twist of luck, "They came off with no other harm than their fright and being very wet, as were all the chests." On Chappe's boat ride toward land, he wrapped up his clock and kept it close. "I . . . sat down upon it myself, to keep it dry in case the waves should chance to wash us," Chappe wrote.

The ocean had already soaked the longboat's passengers on approach to shore. And as breakers pushed the craft toward its uneasy meeting with white sand, the saltwater spray ensured no clothes or unpacked provisions came ashore without a briny overscent coloring the sweat and stench of a long passage. The boom of a vivacious gulf now safely behind him was all the roaring crowd Chappe needed.

"Then it was that casting my eyes upon my instruments that lay all around me, and not one of them damaged in the least," Chappe reflected, "revolving in my mind the vast extent of land and sea that I had so happily compassed, and chiefly reflecting that I had still enough time before me, fully to prepare my intended observation, I felt such a torrent of joy and satisfaction, it is impossible to express, so as to convey an adequate idea of my sensation."

The sun hung low over the scrubby San Felipe foothills to the northwest. Nightfall was too close at hand to venture to San José del Cabo's active mission seven miles inland. Instead, the abandoned former mission at the edge of the beach was their nearest shelter for the night. Freshwater from the nearby lagoon quenched the party, while fresh pitahaya fruit must have tasted like multicolored manna to stale mouths deadened by salt meats and hardtack.

One part of the abandoned beachside property, however, was active. The nearby cemetery kept an informal history of the region told in tombstones—grave markers for local missionaries and converted indigenes of the peninsula, dating back to the mission's founding in 1730.[26]

And judging from the number of fresh graves, in fact, history was still being made. Word had been spreading across the region that for nearly a year a brutal fever had been carrying off both Spanish and native populations like nothing since the spotted fever epidemics of the 1740s.[27] Some called this new plague measles, others a different kind of *grande enfermedad*. All who knew enough to say knew enough to advise the travelers to stay far away from anyone infected.

The thundering surf feeding in from the bay sounded a steady and soothing drone to the travelers who at last lay down for the night, casting their thoughts northward to their final inland destination just a couple of leagues away.

A welcome sleep washed over the voyagers. Even as fits of chills and shivers gripped stricken locals in the epidemic's deadly embrace, a warm offshore breeze kissed the nearby jacaranda trees, cradling their purple blossoms softly to the ground.

Chapter 7

GREAT EXPEDITION

Prague and Central Bohemia
May 1–6, 1768

The window to an unfurnished room slammed shut. From behind it, sharp words muttered in a vaguely Scandinavian-sounding language filtered out into the spring night. The linguist and astronomer Joannes Sajnovics and his boss Father Maximilian Hell—two Jesuit scientists on their way to observe the 1769 Venus transit from a northern Norwegian island—had taken lodgings for the evening in the Bohemian town of Kolín. And Sajnovics was not pleased.

"They barely gave us anything for dinner," Sajnovics (pronounced SHINE-oh-vitch) groused in his travel diary. "We spent the night sleeping on a tiny bit of straw. We could hear the most vacuous music from the neighboring tavern, as much as the impossible sound lacking any refinement of two whistles can be considered music. Good thing they stopped it around ten."[1]

Hell and Sajnovics were Hungarians who had accepted an invitation from the teenage king of Denmark to gather Venus transit data from a desirable location in his kingdom: an island garrison that's practically as close as the European continent gets to the North Pole. Hell, who had met Jean-Baptiste Chappe d'Auteroche on the Frenchman's way through Vienna in 1760, was a Jesuit man of science. Hell enjoyed a place of prestige as chief astronomer to Holy Roman empress Maria Theresa and had edited a lunar calendar and ephemeris that had pre-dated Nevil Maskelyne's *Nautical Almanac*—and was intended more for fellow astronomers' use than for facilitating navigation at sea. Hell had also earned some clout demolishing the claims, some coming from Danish astronomers, that Venus had its own moon. (Hell was right to criticize. The planet Venus, as we know today, does not have any moons.)[2] However, Hell, who'd turned forty-eight in January, was no youngster. And the 1761 transit had left the most important number in astronomy frustratingly indeterminate. The solar distance, also called the astronomical unit (AU), still demanded its celebrated discoverers. The only remaining opportunity the century provided to unveil it still yielded the greatest calling any man of the stars might have hoped for.

Hell had turned down two other propositions to observe the transit from other locations. But the Danish king's offer placed him in an ideal position. The best astronomers in the world, French and British, were observing the transit at tropical locations in the South Seas and at the distant edge of New Spain. But their results would be practically meaningless in the quest for the AU unless they could compare their data with similar observations performed at higher, preferably arctic, latitudes. That a Protestant monarch would invite a Jesuit scholar into his kingdom to gather this needed arctic data could only, as Hell saw it, have been the intervention of divine providence.[3]

Accepting King Christian VII's invitation, Hell picked Sajnovics as his junior scientist, a thirty-five-year-old former assistant from the Vienna observatory who was a fellow Hungarian and Jesuit. Brooding and saturnine, Hell had found a travel mate whose animated and tetchy travel diary entries showcase a pair of men almost comically misfit for one another. Their entourage included a servant, Sebastian Kohl, and Hell's dog, Apropos.

As Sajnovics recorded it, the duo's audience with the empress in Vienna had left the assistant a little starstruck. "Her Highness asked for my name," Sajnovics wrote in a letter at the time. "She asked me where I was from and who my relatives were. Moving on to my astronomical studies, she asked about them so affably, in such a friendly and gracious manner, that her confidential and relaxed conversation completely beguiled me. Finally she made me promise that I would bring back her Hell safe and sound."[4]

Hell and Sajnovics were now making their way north from Vienna to the German port city of Lübeck—where their scientific instruments had already been shipped—whence they would sail to Copenhagen and ultimately embark for their northernmost Scandinavian destination. The voyage carried the familiar clocks, telescopes, and quadrants necessary for precision measurements of Venus crossing the sun's disk on June 3, 1769. But Hell and Sajnovics also brought with them magnetic compasses to satisfy their own scientific curiosity about the subtle variations in the earth's magnetic pull.

Since geomagnetism was a new science, one that some contemporaries were exploring as possibly another means of determining longitude at sea, Hell saw his arctic voyage as an opportunity to explore the frontiers of human knowledge, whatever form it may take.

As Hell would later write in a letter, "Along with these astronomical tasks I shall not neglect work related to the realm of the physical, such as magnetic [measurements], observations with barometers and thermometers, northern lights, and the tides. That means everything I find useful for astronomy, navigation, geography, physics and understanding of nature; all of this will contribute to my work."[5]

Reaching Prague on May 2, Hell and Sajnovics paid their homage to the astronomers at the local university the next day. Having enjoyed a quiet breakfast after morning mass, the travelers ascended the tower housing the college's observatory. ("It is very tiresome getting up to it on the wooden steps," Sajnovics recorded.) The visitors introduced themselves and admired the observatory's instruments. Sajnovics was impressed with the local handiwork behind the astronomical quadrants. "These two instruments were made here in Prague," Sajnovics wrote (but presumably did not utter to his hosts). "So carefully, exquisitely and splendidly that one might consider it an English job."[6]

Despite a day that threatened rain and thunderstorms, Hell and Sajnovics rode in the college rector's four-horse carriage across the Vltava River to visit St. Vitus's Cathedral. "The vicar thought we were Italian abbots," Sajnovics recalled. "We were wearing completely black dresses with a collar—like abbots do—with two hanging white stoles and a small pallium made out of black silk." The pair took mass at the cathedral and kissed the relics of "many famous saints," including that of the medieval martyr who is also the national saint of the Czech people, St. John of Nepomuk. They took mass again the next day, Thursday, May 5, and continued their northward trek toward the land of the midnight sun. Sajnovics, a gourmand at heart, relished his pheasant dinner and trout lunch the next day. The expedition followed the river out of Prague, even as the Vltava becomes the Elbe approaching Germany. "We set out on a wicked road, on the left of which the river Elba was flowing, constantly fearing that rocks would fall off the steep cliffs," Sajnovics recalled. "We spent the night in a lonely house. From here we could see a white chapel on the top of a mountain so high that from afar it seemed to be the biggest mountain of Bohemia, almost impossible to climb by man. That's it for today."[7]

Traventhal House and Surrounding Duchy of Schleswig-Holstein (Today Northern Germany)
May 30–June 1, 1768

By the end of May, the party was approaching its first seaport, Lübeck. Its original intent was to sail from Lübeck to Copenhagen and meet the Danish king who had so graciously enabled the voyage. However, along the way, the travelers learned that the nineteen-year-old recently crowned monarch, Christian VII, had just embarked for a grand tour of England and the Continent. Kings of Denmark at the time also carried the title Duke of Schleswig-Holstein, the region where the visitors now stood. The dukedom's seat lay eighteen miles west of Lübeck. Traventhal House—although the word "house" does meek justice to the pleasure palace it describes—was a beehive of activity as Sajnovics and Hell's carriages pulled into town.

"We arrived in the village of Traventhal in the evening," Sajnovics recorded in his diary for May 30. "Everything was crammed with people. We had to make do with any old room in the inn. After a bit of soup and some beer, we spent the night on hay, as the other beds were taken by the owner and his staff. It was a miserable night. We found out that the king had arrived a few hours before at the palace."[8]

Just the day before, the king had placed a capstone on a sly diplomatic victory for Denmark—a victory that also represented one of the greatest strategic blunders in the career of Russian empress Catherine the Great. Catherine's husband, Peter III, had six years earlier been advocating an ancestral claim to the very same Schleswig-Holstein dukedom where Hell and Sajnovics had arrived. The dukedom, crucially, would have given Russia the temperate seaports it needed to become a maritime power to rival at least the Dutch if not also the French and British. However, Catherine viewed her husband as a thwart to her own ambitions, so once Peter had been shuffled off to the great toy land beyond, Catherine papered over possible points of contention with any of her Teutonic countrymen. Emperor Peter's saber-rattling plans for the future of Russian sea power became, under Danish suasion and Catherine's envoys, a humbler request to endow less strategic German lands upon members of her extended family.[9]

On May 29, King Christian VII granted earldoms to key players in the Russian negotiations. One such entitled gentleman was Denmark's brilliant minister of foreign affairs, Johan Hartvig Ernst von Bernstorff.

Upon arriving at Traventhal on May 31, where the Hungarian visitors would be meeting their sponsoring king, Sajnovics met Count Bernstorff. Not only was the foreign minister well-known in Denmark, but courts across Europe also recognized Bernstorff's preeminence. Prussia's Frederick the Great—who called him "the oracle of Denmark"—reputedly quipped, "Denmark has her fleet and her Bernstorff."[10]

Sajnovics was clearly impressed. "This prestigious minister is nearly 45 years old, of medium height, with a friendly disposition, a soothing voice, very gentle, an exceptionally intelligent and extremely cultured man," Sajnovics wrote. "I could never praise highly enough the friendliness, kindness, respect and admiration with which he received Father Hell. He said he was happy to see such a meritorious man visit his entire country and that he would try to make his stay in Copenhagen and Vardø [the transit expedition's ultimate destination] as pleasant as it deserved on account of his extraordinary merits."[11]

Taking a quiet lunch at a nearby inn, Sajnovics and Hell were surprised to see one of the king's counselors approach their table. His Majesty, the envoy said, requests your presence at lunch at the palace. The king, they learned, took lunch at four o'clock.

"The lunch was sumptuous," Sajnovics wrote. After the afternoon repast, during which they had no opportunity actually to meet the king, Sajnovics and Hell took a walk among the famous sculpted greenery surrounding Traventhal House. "Walking about in the wonderful garden created by man and God, we had the marvelous idea to have a look at the maze at the end of the garden," the visitor recorded.

A river flowed through the lush baroque landscaping, and Sajnovics began to wonder at a small boat that first approached them and then, once the travelers had passed, turned around as if to track them. "Wer sind Sie?" ("Who are they/you?") asked the boat's sole occupant, wearing a hood over his face. "Since we did not know who was asking, we just ignored the question," Sajnovics wrote, "but he repeated the question louder still. Wer sind Sie? At the same time, he grabbed the paddle to push the boat that had been stuck on the shore back in the water. The silver cross on his chest, the collars of the Order of the Elephant and all of his appearance indicated that he was the king!"

Hell genuflected and immediately apologized for not recognizing His Majesty. But the king would have no such formalities once he'd learned who the two gentlemen were. "Pater Hell?" King Christian VII replied in German—the lingua franca of the Danish court. "Kommen Sie zu mir!" As commanded, Father Hell approached the king and a brief exchange of greetings passed their lips. However, without Bernstorff or other advisers at hand, the king could not accept an official visit. Instead, he requested his guests also attend dinner—which Sajnovics later learned began at 10:00 PM or later. In no mean feat of cheek, Hell and Sajnovics declined the invitation and showed up at the palace the next morning instead.

Sajnovics and Hell arrived at Traventhal at 11:00 AM. (Both days they'd left Apropos behind with the innkeeper, "growling and protesting," Sajnovics said.)[12] Talking both courtly and scientific matters with the royal advisers, the pair learned the nineteen-year-old monarch would greet them in person at 4:30 PM.

The appointed hour came, as did the next hour, and the hour after that—without a hint of a royal audience. "Around seven o'clock the king left the meeting room," Sajnovics recorded. "His purple dress was embellished with silver . . . a wide dark blue royal belt on his waist, a sword on his side, and his hat held under his arm. He is of medium height, proportionately built; he has an open and gentle look, only 22 years old [*sic*]; his royal dignity is doubled by his handsomeness, so anyone who gazes upon him cannot help but love him. We greeted him . . . and he expressed his gratitude in German that P. Hell had safely arrived and said that he was very happy to have received such a great astronomer to carry out the astronomical observations in his country."

Christian queried Hell about their backup plans in case of inclement weather, as well as their expedition's other (nonastronomical) observations. All depends on the will of God Almighty, Hell replied. "Then the king turned to me and asked whether I was also a mathematician," Sajnovics wrote. "I have studied the same sciences—I answered—and I am at His Majesty's humble service."

The three-way conversation continued for another thirty minutes, including the king's observations on the voyage's destination. "[We talked] about the fact that he would send a biologist along with us from Copenhagen and other such things," Sajnovics said. "We departed kissing his hand. He returned to his chambers."

Hell found His Majesty still more impressive than did his assistant. "I myself was greatly amazed by the regal qualities of the [king's] soul and body during the three private conversations I had with him in Traventhal, two of which were exceptionally long," Hell later wrote in a letter to Father Höller, empress Maria Theresa's confessor. "And I must admit: Because of these qualities I developed such an affection and respect for this young king that no matter where I was traveling I considered those royal subjects who have been blessed by God with such a great king most happy."[13]

Sharing his observations about a foreign king with such a prominent figure in Maria Theresa's entourage—someone who might share Hell's remarks with others at court—Hell was not in any position to be candid. In fact, whether Hell noticed it or not, King Christian VII was a sick man. Suffering from a degenerative psychotic condition (which later ages would diagnose as dementia praecox), the king was coming through Traventhal as one of the first stops on his grand tour.[14] The king's advisers hoped the trip would be the curative that could rid His Majesty of the mental breakdowns that everyone at court had been keeping under wraps. Married to his cousin, the sister of England's King George III (himself no stranger to bugbears of the brain), the young Danish king already had a reputation for inciting scandal during his two years on the throne. Chief among them were the sadomasochistic encounters he was reputed to have had with one or more confidantes at court, as well as a Copenhagen prostitute—Støvlet-Katrine ("Boots Catherine")—who had become the king's official mistress.[15]

Earlier, on May 29, the Danish king had discharged his court-appointed doctor, leaving the unstable monarch without any medical or psychiatric supervision. The English ambassador to Denmark later quipped that "the politics of Denmark seem incapable of any other production than that of intrigue."[16] As later events proved, the man hired to take the place of King Christian's dismissed physician would reduce the diplomat's observation to a feeble understatement.

PILEGRIMSLEDEN (PILGRIM'S ROUTE) FROM OSLO TO TRONDHEIM, NORWAY
July 22–August 22, 1768

Roughly tracing a centuries-old pilgrimage road for early Christians visiting the tomb of St. Olaf, the "Eternal King of Norway," the transit pilgrims Hell and Sajnovics had continued their apprenticeship in the art of roadway improvisation. A troublesome axle on one of their carriages kept breaking, and vertiginous cliff roads and rushing rivers had forced the caravan to make snap decisions that kept these men of God praying. "The axle was grinding against the cliff wall on the right, which was about fifty feet high vertically and in fact so much bent towards us that we were in constant fear of the entire mountain collapsing on our heads," Sajnovics wrote on July 22 on a narrow mountain road outside of Løsnes, thirty miles north of Lillehammer. "Two beam lines on the left are meant to ensure that the slightly wayward wheel does not cause the carriage to fall into the dreadful abyss."[17]

The contrast between the group's royal trappings and the decidedly rugged terrain was something to marvel at. Father Hell rode alone in a red-and-gold-trimmed single-seater sprung carriage; Sajnovics's ride was similarly ornamented but painted in green. The baggage—and servants—tailed behind in five separate dual-horse-drawn carriages. "The mountains resemble big cones that are empty below and shaped like big ovens, among which the sound of the carriage wheels imitates the rumbling of the thundering sky," Sajnovics later wrote to his Hungarian Father Superior. "Imagine a gigantic mountain which has vertical cliff walls on the side; over your head the threat of an imminent fall of rocks and under your feet flow the waves of mountain creeks among peaks of sharp cliffs of hair-raising depths, the sight of a dark and deep lake, or the bay of a waving sea gaping at you from down below. And you must go along this cliff wall on a road so narrow that if you turn a little to the left you cannot avoid the danger and would almost certainly lose your life. Yet the overhanging cliff sometimes does make you move to the left! Ah, if only the Viennese could have seen us . . . in these desperate moments, they would have forever given up the hope of ever seeing us again."[18]

Yet one day's rosary-bead clutching sometimes cued the next day's moments of jaw-dropping awe. On July 23, Sanjovics marveled, "Today's journey was the most beautiful by far. The wide plain was accompanied by mountains on both sides, reaching far up high, where they were covered by woods. The valley was crossed by a sea-blue river abounding in fish. . . . Part of the population was scything the beautiful grass, which was then put up to dry on horizontal bars hoisted onto firmly grounded poles; others were enjoying the harvesting month, made promising by the heavy wheat. . . . The waters gushing down from the cliffs and the mountains, mostly vertically, offer a spectacular view for someone who is not used to such sights."[19]

Hell had hired a servant to travel ahead of the group by a half day, anticipating the caravan's quotidian needs—ordering lunches and dinners at inns, arranging for fresh horses to be available wherever possible. ("We needed 4 for the carriage[s], 6 or 8 for the carts, one steed for the servant and also one for the Danish interpreter who joined us in Copenhagen," Sajnovics later recalled.)[20] Because the Norwegian Sea grew increasingly treacherous as the darkness of autumn descended, the plan was to arrive at Vardø, their ultimate destination, in August and spend the entire fall, winter, and spring setting up their observatory and making astronomical, magnetic, and meteorological measurements. But as the summer's purple heather began to disappear from the roadside's tender color palate, a storm-soaked autumnal passage to Vardø looked increasingly likely.

Showers on land were no great joy either. But for these men, rain delivered in quantity meant not just seeking shelter but also uncasing various instruments for scientific measurements. On July 25, in the midst of the Dovrefjell mountain range, lightning bolts illuminated a darkened afternoon sky as Hell's magnetic compasses deviated 19 degrees westward from magnetic north—presumably due to the storm itself. Earlier in the day, under clearer skies, the explorers had also used a quadrant on loan from the German explorer Carsten Niebuhr to find the sun was 42 degrees, 16 arc minutes, and 50 arc seconds from the horizon. Reminders of winter remained underfoot, even on a summer's day. "We were laughing that on July 25 we were able to walk on snow and touch it," Sajnovics wrote.[21]

Good spirits kept the group going, as a few days later an accident would provoke the same response as the snow. "The wheel bumped into a cliff, causing the carriage to capsize and everything in it spilled out," Sajnovics later wrote. "Just imagine: bacon, bread, boxes, and wine flasks, etc. lying all over the place. We could not help but laugh about it."[22]

Tight ravines cradled rocky torrents as the travelers wound their way northeast. They were only four days away from Trondheim, the Norwegian port city that would serve as their way station before shipping out to the arctic. Four days is a long time, though, when nature seems fixed on holding back every forward step.

"We were going over a sea of cliffs, along a rushing river crossed at many points by strong bridges built into the cliffs," Sajnovics wrote at the end of what he figured the expedition's worst day, July 26.[23] He later described the alpine passage in a letter: "These mountains always look like they are splitting into two. When they do, the two sides are connected by bridges. But what bridges! They are merely wooden planks without anything securing them. The minute the horse steps on it the whole thing starts swaying and the foreigner begins to be frightened thinking that he might find his grave among the rushing waves down below. In such dangerous traveling conditions, it is no wonder that our carriages were always breaking. Either the axle cracked or the wheels. Mending them always cost a lot of money and time."[24]

On July 30 the battered train of broken royal carriages and baggage carts pulled into Trondheim. Sajnovics and Hell returned—one last time—to the luxury they had come to know as the guests of King Christian VII. "We stayed at the inn that had already been ordered for us by the governor of the region," Sajnovics wrote back from Trondheim to his Hungarian superior. "The town officials came one after the other to pay their respects to Father Hell." The Hungarian travelers had seen similar pomp on their earlier visits to Oslo ("received by the officials of the town in the company of all the nobility in great festivity and with admiration") and Copenhagen ("It is almost unbelievable how they took to us; the scientists all feel honored to be able to seat us at their tables … everyone is keen on meeting the brilliant Hell face-to-face and talk to the man known by his works and reputation; we are truly wonder-men here").[25]

A third scientist joined the expedition in Trondheim, Jens Finne Borchgrevink—a man who knew the local dialects, had traveled in the region before, and even had family in the area.[26] Borchgrevink was a student of the legendary Swedish botanist Carl Linnaeus. Sajnovics began addressing his botanist colleague as "the student," although the man was thirty-one years old. "He is a student of the nature of sea plants and algae, and also an initiated botanist, having attended the courses of Linnaeus in Sweden for a whole year," Sajnovics recorded in his diary. "He thought he might learn a thing or two about the astronomical sciences."[27]

Sajnovics, Hell, and Borchgrevink spent much of August awaiting a ship to be readied for the dangerous passage north. Hell hired a mariner named Friedlieb to outfit a *jagt* (yacht) for the sea voyage to Vardø. "He ordered Mr. Friedlieb not only in his name but also, in the name of the king, that P. Hell's expenses must be covered no matter what he wanted or where he was going, in accordance with the royal ordinance sent by the King to Trondheim," Sajnovics recorded.[28]

Passing the time, the Hungarian visitors wowed town officials with tricks using Sajnovics's and Hell's magnets. Hell gave Catholic mass to penitent Danish soldiers—many of whom were hired hands from papist countries in Europe—garrisoned nearby. One of their traveling companions, son of the town's late bishop, showed off his musical talents. "Mr. [Eileru] Hagerup plays the piano and the flute and he can sing," Sajnovics recorded about a concert the town gave for Hell and Sajnovics on August 2.[29]

On Sunday, August 21, Hell and Sajnovics held their last formal mass and began packing for the journey north. Their *jagt* was ready. Originally named *Anden* (Norwegian for "The Duck"), the boat was rechristened *Urania*—after the ancient Greek muse of astronomy.[30] The town governor had, Sajnovics wrote, "equipped a large ship for us that was big enough for all our equipments and could also resist the waves of the roaring sea, hiring five sailors who were to take us there, spend the winter with us on the island and bring us back the next year."

Their winter needs—wood and coal as well as fish skin gloves and every fur and wool garment an arctic adventurer could want—were well met. A cook and pastry chef shipped out with them too, Sajnovics continued, complete with a hold well stocked with "meat, salted fish, bacon, peas, beans, carrots, cabbage, two receptacles filled with large quantities of flour. Moreover, they sent tea, coffee and Dutch chocolate, as well as cutlery for all of these. . . . They did not forget about wine either. We had red and white wine at our disposal; more than that, they also gave us brandy, and all the necessary tools to brew beer."[31]

At noon the next day, the *Urania* set sail with twelve men aboard—the three scientists, one gentleman passenger (Hagerup), two servants, plus the cook and five crew.[32] A storm descended on the ship within four hours of launching, forcing it to anchor at the nearest port, Lensvik. "No sooner did we throw in the anchor than a terribly vertiginous wind started around six," Sajnovics wrote in his journal. "There were plenty of fish around here too. We could hear them jumping and playing in the water until late at night. We had fried fish around ten, washed down with red wine. Sleep."[33]

The Norwegian Sea, en Route to Vardø
August 30–October 8, 1768

Nautical adventure books of the day told of sleek French frigates and early English clippers that could zip across stretches of the Atlantic and Indian oceans at almost incomprehensible speeds. The *Urania* was not one of those ships. She had just one sail, albeit a multipurpose sail with detachable sections that could billow deep in accommodating winds but also trim light when storms raged. The ship spent the rest of August covering nautical mileage that could have been bested on land by oxcart.

For three August nights, with only headwinds to face, the group whittled away the hours in a tavern in the town of Vallersund, with Father Hell plinking out songs on a new mandolin and the party's gentleman guest, Hagerup, accompanying on flute and dulcimer.

The month's final two days could well have ended the whole expedition. "We had to sail on a famously dangerous eight mile portion of the sea, known for the many shipwrecks, situated far away from the shores," Sajnovics later recalled in a letter. "Our ship was lifted as high as on a mountain by the ever-growing waves and then it seemed to be falling back to the bottom of the sea, and then it was being thrown around to the left and right, threatening to fall apart among creaking sounds. The rainy fog was making the situation even more dangerous, because we could not see the mountains nor the rocks in the sea."[34]

Much of September bore down like the headwinds that increasingly battered the *Urania* and forced her in to port for five days and four nights of waiting for favorable weather. On two different days that month, the travelers passed the time in "miserable expectation," as Sajnovics wrote, by collecting local sea life onshore. On the third, for instance, some hired hands helped them dig for clams. Sajnovics noted that the locals "were eating the clams in Normand style, raw and with great appetite," while the visiting scientists watched "with eyes and mouths wide open—but still would not join them." Instead, they ate lunch and dinner in the pouring rain, whipped by a bitter northern wind.[35]

Other exotic life-forms in and around the Norwegian Sea had long been rumored. And while the scientists on the expedition did not bother collecting such tales, crew members on the *Urania* had no doubt been well schooled in the local myths. Mermen, mermaids, monstrous sea snakes, and the dreaded kraken were all reputed to live in the waters off the northern Norwegian coast. Some almanacs of the day spoke of these beasts matter-of-factly, as if rattling off peculiarities of the climate or local cuisine. One gazetteer, for instance, wrote that Norwegian mermen and mermaids "are so well authenticated that I make no doubt a new and very surprising theory of aquatic animals may in time be formed."[36] And the kraken—a behemoth octopus-like sea creature that could allegedly swallow entire ships whole—had practically earned the title of most fearsome beast on the planet. "The most surprising creature in this sea, and perhaps in the whole world, is the kraken, or korven, an animal of the polypus kind, seeming a mile-and-a-half in circumference," a popular English author wrote in 1768. "The Norwegian fishermen sometimes . . . know the kraken is below them, and that they are fishing on his back. When they perceive, by their lines, that the water grows more and more shallow, they judge he is rising slowly to the surface and row away with great expedition."[37]

Yet not every legend in these storied waters went unobserved. The Venus transit voyagers spent September 25–26 squaring off against the Mosktraumen—the famous oceanic vortex that in the coming century Edgar Allen Poe would immortalize as the "maelström," and Jules Verne would set as the final destination for his fictional submarine in *20,000 Leagues Under the Sea*. Both nineteenth-century authors turned what was a dramatic—and potentially deadly—force of nature into the stuff of melodrama and hyperbole. Sajnovics, by contrast, simply chronicled what he saw.

"We got close to the island of Stromoe [Ryøya] having sailed between mountains covered in snow," Sajnovics wrote. "On the edges of the gulfs the water was wrathfully ruffling, and it was flapping incredibly hard, speeding up the twirls of each and other. . . . In this section the water seems to be still, compared to ones around it, where the frequent waves that are not big, but very high and steep, rush and roll very quickly. Many fishing boats are lost here every year."[38]

Yet September was just a teaser for attractions still to come.

On October 1, the *Urania* crossed into the waters off Finnmark, the northernmost county of Norway, where Vardø was. "On the 2nd we could start again with a favorable wind," Sajnovics later wrote. "But because of the thick snow that was covering everything in sight we had to stretch out the entire sail so that we would not hit the cliffs, and so we were exposed to the rage of the wind and the storm to such a degree that the captain nearly got blown away from the steering wheel."[39]

Sleet storms, fierce tides, and changeable gales had conspired to lead the *Urania* into the rocky shallows. Three days of repairing the ship followed, with the next day on the water nearly dashing them against the rocks once more. "We were facing the most dangerous portion of our journey thus far," Sajnovics journaled. "We had to cross two big fjords, the deep Eastern Sea under us throughout the whole time. Both fjords were being perturbed by gigantic waves, and if the wind happened to change direction, these waves would have brought on an even greater danger for us."[40]

But that night, the seas calmed down enough to permit luminous plankton to cast an eerie green pall across the *Urania's* battered hull. "The marine phosphorescence gave us a surprising show," Sajnovics recalled. "It was like our ship was cutting through fiery foams."

On October 8, the *Urania* reached the tiny settlement of Hamningberg—less than fifteen miles from Vardø. But unforgiving winds and an unaccommodating shoreline left the boat with practically nothing to secure it to land. "We threw in three anchors waiting terrified for what fate would bring next," Sajnovics wrote. "The wind was raging terribly throughout the night and we were expecting the ropes to be torn at any minute. But lo and behold, we saw the dawn! The two main anchors had fallen on cliffs, so they barely held anything. But one that was most worthless sunk in close to the shore, and it was the only one that resisted the attacks of the [waves] and the winds."

Vardø, Norway
October 11–November 20, 1768

Any European of the age who enjoyed adventures, hazards, and perils could scarcely go wrong buying a book about the arctic. Readers vicariously venturing into the far north could expect to discover "the bleak and chilling prospects in the Arctic seas" or that "les vents y sont en toute saison d'une impétuosité qui rend la navigation très-périlleuse."[41] ("The winds are so forceful in every season that they make travel by ship very dangerous.")

Yet as Sajnovics discovered, the dangers now facing his expedition had drawn themselves out into something so continually mortally humbling that the experience couldn't quite be expressed in words.

One final storm front followed *Urania* into Vardø, and this gale too nearly sunk the boat. "Only he who knows the sea can know the dangers of the sea," Sajnovics later wrote in a letter. "There was one enormous wave that was angrily chasing our ship, splashed above it, and . . . in its anger it hit the cabin, penetrating through the windows and the cracks, it filled up the whole room." Fortunately, angry seas also meant swift waters, so *Urania* arrived in Vardø "in two short hours that seemed very long," as Sajnovics noted. "No sooner did we get off the ship than the wind started screaming again, the sky darkened and a veritable tempest started that kept raging for a whole day; they all agreed that we would have perished, had the tempest caught us at sea. Beg the Lord to keep us safe the next year too."

Little wonder, then, that the military officials manning the remote Norwegian garrison at Vardø were surprised when Hell and Sajnovics arrived. Sailing in mid-October through the Barents Sea was a calling that fell somewhere on the recklessness spectrum between crazy and suicidal. Nature's fury in the stormy season was, as Sajnovics had seen firsthand, something bordering on unnatural. And as nature's annual payment for being the summertime land of the midnight sun, daylight would be quitting the region in a month.

Hell and Sajnovics saw their safe deliverance as the kind of beneficence that called for deep reflection and prayer. "We set up the altar in the neighboring small room, and we expressed our gratitude to God," Sajnovics wrote on October 12, the day after their arrival in Vardø. "We received visits from [Israel Olai Sigholt] the commander, [Peder Fischer] the lieutenant, [Raskvitz] the exile, and [Voigt] the barber—who did not actually know how to cut hair."

A mostly torch-lit procession of trips back and forth to the ship followed. And out of the snow, in stakes and twine, emerged the beginnings of the observatory that the two astronomers had already sketched out. Their new building, near the village center, would be an annex to a house of one of the local officials, the *Fogdens hus*.[42] The next day,

Sajnovics and Hell took lunch at the vicarage. "The sky was covered in snow clouds," Sajnovics recorded. "Our hunger and thirst were amply settled with noodles, lamb stew, and Schneeballen red [ice] wine." Table talk centered around the poor fishing season that had just ended and the outbreak of scurvy that Vardø had endured the last time such meager stocks supplied the long winter. So during the coming darkness, reindeer meat would be the new staple.[43]

Days of porting chests and timber through bracing winds gave way to insomniac nights kept restless by mice chewing on the visitors' lodgings. Hell and Sajnovics installed their Niebuhr quadrant—the one they'd already tested on the road to Trondheim—and measured the sun's altitude through intermittent breaks in the clouds. On October 13, Hell ordered a day's expedition to a nearby island to collect moss that would serve as the observatory's insulation. The next day, through a thick layer of snow, they laid the stone foundation.

On October 27 and 28, a raging arctic blizzard battered the island, blowing bitter winds through the cracks in the walls and keeping the shivering travelers up most of the night. Beer and water stores froze and burst their containers. "We took the wine to the pantry of the commander, lest it suffer the same fate," Sajnovics wrote.

Some of the men were still quartered on the *Urania*, anchored near shore. Fierce, towering whitecaps crashed onto the ship and tossed it around like a toy boat. "Had it not been for the rope holding it down on the other shore, the wind would have definitely thrown our ship out into the open sea together with the two sailors sleeping on it," Sajnovics wrote. "In the evening the waves became strong again, so the captain made the soldiers pull the ship closer to land."

Those measures weren't enough. A fortnight later, on the night of November 8, Sajnovics noted that "the remarkably big waves had thrown our ship against the shore, having ripped apart the ropes that were used to secure it. It was filled with water. The captain was crying and whining." [44]

The days grew progressively shorter and dimmer. On November 18, when the sun rose at 11:32 AM and set at 2:06 PM, three soldiers and a peasant from town set out in the captain's boat to fish.[45] "Suddenly," Sajnovics recorded, "an unexpectedly strong east wind started blowing and raved with fury throughout the day, to such an extent that we lost all hope of ever seeing those men again."

Although the fishermen miraculously survived the night and made it back to shore the next morning, their return also marked a dark day on the local calendar. November 19 was to be the last day of sunlight for 1768.

"When we saw the sun rise in the south on 20 November and soon after go down in the same place, at that point it said good bye to us for a very long time," Sajnovics later wrote to his Father Superior. "The light was replaced by darkness, indeed a fabulous and Egyptian darkness."[46]

Jean-Baptiste Chappe d'Auteroche (b. 1728) was a French polymath astronomer with a keen sense of wonder, a flair for drama, and an instinct for cutting-edge science. Chappe's observations of the transit of Venus in 1761 and 1769 provided some of the best data to answer one of the age's greatest scientific problems: How far away is the sun? *Engraving from* Voyage en Californie, pour l'observation du passage de Vénus sur le disque du solei *by Jean-Baptiste Chappe d'Auteroche (1772)*

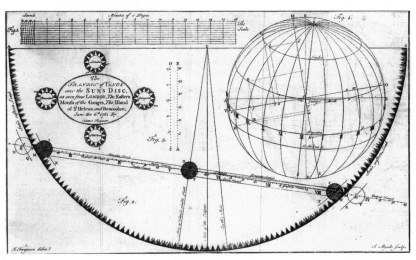

The Scottish astronomer and instrument-maker James Ferguson presented a series of lectures in early 1761 that projected the path of the planet Venus as it would be crossing the sun on June 6 of that year. Through nineteen pages of complex geometric calculations, Ferguson offered his readers, he wrote, a way "to trace this affair through all its intricacies [but] to render it as intelligible to the reader as I can." *Diagram from* Astronomy Explained *by James Ferguson (1764)*

On his Venus transit voyages to Siberia and present-day Mexico, Chappe took regular observations of everything from ocean currents to atmospheric pressure. Here he delights at lightning discharges during a thunderstorm while his less enraptured servants and Russian hosts take shelter. *Engraving from* Voyage en Siberie, fait par ordre du roi en 1761 *by Jean-Baptiste Chappe d'Auteroche (1768)*

In 1763, the British ship's master Archibald Hamilton wrote out a manuscript account of his ocean travels. On the cover page he drew this image of a master's duties—tracking positions of sun, moon, and stars with his "quadrant" and chronicling it and other data in the ship's log. *Courtesy of the National Maritime Museum*

The Rev. Nevil Maskelyne (1732–1811) was an astronomer and fierce advocate of navigation at sea via "lunars"—longitudes determined by the moon. Maskelyne was Astronomer Royal of England from 1765–1811. *Smithsonian Museum*

Naval Lieutenant James Cook, later Captain Cook, headed up a Venus transit expedition to Tahiti that set sail from England in August 1768. Cook commanded a refitted collier called *Endeavour*. *National Library of Australia*

Key figures in the *Endeavour*'s 1769 Venus transit expedition to Tahiti: (l. to r.) supernumerary Joseph Banks; former Lord of the Admiralty, John Montagu, Earl of Sandwich; Lieutenant James Cook; botanist Daniel Solander; and chronicler John Hawkesworth. *National Library of Australia*

The *Endeavour* in Australian waters in 1770. "A better ship for such a service," her captain James Cook wrote, "I never could wish for." *National Library of Australia*

The Hungarian Jesuit father Maximilian Hell, who along with his assistant Joannes Sajnovics, traveled to an island in northernmost Denmark (today Vardø, Norway) to observe the 1769 Venus transit. *Engraving by Johann Gottfried Haid from a painting by Wenzel Pohl*

Endeavour's artist Sydney Parkinson made a thousand drawings during the ship's voyage, including sketches of the encampment that Cook had built at Tahiti for Venus transit observations. This engraving is based on Parkinson's "Fort Venus" sketches. (Parkinson himself died of dysentery on the journey home.) *National Library of Australia*

A former Jesuit mission outside the town of San Jose del Cabo (in present-day Mexico) served as the site of Chappe's 1769 Venus transit observation. An epidemic of typhus was also decimating the region at the time of Chappe's journey here. *Courtesy of Musée du Louvre, Département des Arts graphiques*

Chapter 8

SOME UNFREQUENTED PART

MADEIRA ISLAND
September 12–19, 1768

The harsh light of a cloudless morning brightened the palette of tan, purple, and greenish tones to the northwest. Britain's HM Bark *Endeavour* pulled round a hilly landmass to meet her maiden voyage's first port of call—the Portuguese-occupied Madeira, an island of cliffs and pinnacles that shoots up out of the ocean like a tiny Dover.

Chief astronomer Charles Green had manned his sextant on *Endeavour*'s approach, measuring the angular altitude of the sun throughout the day and concluding their latitude was 32 degrees and 42 arc minutes.[1] "When you first approach [Madeira] from seaward it has a very beautiful appearance," the *Endeavour*'s gentleman naturalist Joseph Banks recorded in his diary. "The sides of the hills being entirely covered with vineyards almost as high as the eye can distinguish, which make a constant appearance of verdure—tho at this time nothing but the vines remain'd green, the grass and herbs being entirely burnt up except near the sides of the rills of water by which the vines are water'd."[2]

As soon as *Endeavour* had been authorized to anchor in Funchal bay and conduct its business onshore, Lieutenant James Cook—supernumeraries onboard called him "captain"—authorized Banks and his assistant botanist Daniel Carl Solander to land and represent the ship to Madeira's English consul. The two were eager to explore the island, having discovered new species of both marine and bird life during the ship's approach.

Cook had his own business to attend to onboard. The night of their arrival, September 12, the rope attached to the stern anchor had slipped, unsteadying the ship's mooring. At 6:00 AM the next morning *Endeavour*'s quartermaster Alex Weir launched a boat to help raise the anchor, refasten it, and throw it back in the water. As the ship's gunner Stephen Forwood wrote in his diary, Weir's party had "hove [the anchor] up and carried it out again to the eastward, where Mr. Weir's mate, having charge of the boat, by heaving the anchor out of the boat, got fore of the buoy rope."[3] As Sydney Parkinson, the ship's artist—hired by Banks to sketch the many specimens and natural settings on the journey—recorded, "The buoy rope happen[ed] to entangle one of [Weir's] legs, he was drawn overboard and drowned before we could lend him any assistance."

A blanket of clouds and rain descended on the afternoon, darkening the mood appropriately for the first death of an *Endeavour* crewman. Typically a dead crew member's mess mates—the four- to six-man unit assigned to take their meals together—prepared the body for burial. The men wrapped their mess mate's corpse in a patch of canvas, drawn from the material used to mend sails, and threw in two cannonballs to send the cloth casket straight to the bottom. The sheet was sewn together, with the final stitch passing straight through the deceased's nose—both to ensure the man was truly dead and to keep the body fastened to its shroud. The next day, a pinnace lowered into the water, bearing a detachment that would drop the late quartermaster overboard. "[We] sent the boat into the offing," Green recorded. "To bury the body of Mr. Weir which we had found entangled in the buoy rope of the kedge anchor."[4]

Onshore, Banks and Solander had taken lodgings with one W. Cheap, British consul to Madeira. In spite of the late season, after which many flowers had already blossomed for the year, the two naturalists collected 246 plant specimens as well as 18 types of fish. They watched the locals prepare wine from the local vineyards—and were unimpressed. The method, Banks recorded, "is perfectly simple and unimprov'd." Vineyard workers, after removing their stockings and jackets, jumped into vats with the grapes and stomped around. The pulp, Banks said, was then "put under a square piece of wood which is press'd down by a lever, to the other end of which is fastened a stone that may be rais'd up at pleasure by a screw. By this way and this only they make their wine, and by this way probably Noah made his when he had newly planted the first vineyard after the general destruction of mankind and their arts—tho it is not impossible that he might have used a better [way], if he remembered the ways he had seen us'd before the flood."[5]

Despite Banks's disapproval of Madeiran viticulture, Cook sent *Endeavour*'s casks to shore to be filled with water and with 3,020 gallons of the island's famous wine.[6] Madeira wine was well suited for a long-haul voyage because brandy was added to the final product, upping both alcohol content and shelf life.

Cook also ordered his storeroom to be stocked with, as he wrote in his captain's journal, "fresh beef and greens for the ship's company." During the Madeira stay, Cook even had a seaman and a marine whipped for "refusing to take their allowance of fresh beef." Cook also loaded up with fresh fruits and onions. Every man on *Endeavour* was issued thirty pounds of Madeira onions—and expected to incorporate it into his daily rations.[7]

Scurvy had long laid waste to ship's crews at sea. And while no one at the time knew exactly what caused the disease, Cook would be testing new foods throughout his three-year circumnavigation that were thought to prevent scurvy. Such prophylaxes included sauerkraut, a lemon-orange concentrate, and water boiled with a sticky brown bouillon—"portable soup"—made from beef offal, salt, and vegetable stock. Before advancing to the newer curatives, though, Cook would first be fouling the stench of every *Endeavour* crew member's breath. As one treatise on scurvy at the time advised, "Every common sailor ought to lay in a stock of onions, for they are a great preservative at sea."[8]

Unburdened by officers' duties, Banks and Solander had free rein on the island for nearly five days. They ventured into the hills outside of Funchal to visit a doctor who gave them samples of the island's guava, pineapple, mango, banana, and cinnamon tree bark. Another day took them into a Franciscan convent, where Banks recorded his observations on the local climate, population, architecture, and religious culture. "The churches have an abundance of ornaments, chiefly bad pictures and fig-ures of their favorite saints in lac'd clothes," he wrote. The duo visited a nearby convent, where, Banks said, "the ladies did us the honor to ex-press great pleasure in seeing us there. They had heard that we were great philosophers, and expected much from us." The guests regaled the sisters, by their request, with scientific theories about thunder and ways to divine new sources of water. The exchange went both ways too. "While we stayed," Banks wrote, "I am sure there was not the fraction of a second in which their tongues did not go at an uncommonly nimble rate."[9]

On one day the island's governor waylaid Banks and Solander at their quarters, while he talked their ears off. Like many gentlemen of his age, the governor enjoyed keeping current with the latest scientific technology—no small sample of which Banks and Solander had in their possession. The governor's fascination centered around an "electrical ma-chine" built by the London instrument maker Jesse Ramsden. It was an eight-inch-diameter glass disc that was spun as it pressed against a leather pad, generating little lightning discharges from an attached metal rod. With learned organizations around the world, like London's Royal

Society, providing public demonstrations of spark-producing whirligigs like Ramsden's device, polite society produced its own quiet hum about the zaps and jolts such machines produced.[10] Researchers such as Benjamin Franklin, William Watson, and Henry Cavendish had each recently advanced the field with discoveries about lightning, circuits, and conductors. Journals and books enthused over possible medical uses of "the electric fluid." One prominent London book published the year before *Endeavour*'s launch, for instance, detailed numerous cases in which "medical electricity" had allegedly helped patients with their ailments. "It has seldom failed," one account read, "to cure rigidities or a wasting of the muscles—and hysterical disorders, particularly if they be attended with coldness of the feet."[11]

Because of the governor's visit, Banks noted, "we were obliged to stay at home, so that unsought honor lost us very near the whole day. . . . We however contriv'd to revenge ourselves upon his excellency by an Electrical Machine which we had on board. Upon his expressing a desire to see it, we sent for it ashore—and shock'd him full as much as he chose."[12]

NEAR THE EQUATOR, ATLANTIC OCEAN
October 1–26, 1768

Captain Cook and his crew and supernumeraries were not voyaging across the globe just to discover the solar system's dimensions. They'd also been commissioned to test Britain's most celebrated navigational breakthrough: Nevil Maskelyne's *Nautical Almanac*. Together with four human computers (including Charles Green's new brother-in-law William Wales) and the *Almanac*'s "corrector" Richard Dunthorne, Maskelyne had completed a Herculean task.

The debut *Almanac* of 1767—tabulating a stunning 15,500 solar, lunar, planetary, and stellar positional predictions—had been a midnight oil burner, falling far behind its publishing schedule and coming into print six days into the year it was forecasting. A thousand copies had been printed, which promptly found their way onto ships traversing the planet and into navigational schools in the United Kingdom, North America, and Europe.[13]

His book was stunning, but now Maskelyne faced the difficult task of bottling his lightning. How to repeat the same impossible labor year in and year out?

Having been satisfied that the team was indeed up to the job, the Board of Longitude authorized Maskelyne to spend whatever he felt he needed to produce almanacs for 1768 and 1769 and, in due course, into the 1770s as well. So Maskelyne hired more computers.[14] The Astronomer Royal furthermore rewarded innovation among his team members, giving £50 bonuses to computer Israel Lyons and corrector Dunthorne for finding shortcuts in calculating tables concerning atmospheric phenomena like refraction.[15]

The board wanted the *Endeavour* to sail with almanacs for the whole of its projected voyage. Good as they were, Maskelyne's computers hadn't yet refined their methods well enough to forecast so far into the future. They did, however, provide Cook and Green with almanacs for 1768 and 1769. Green was becoming very familiar with them as the *Endeavour* followed the northeast trade winds down to the equator.

On Saturday, October 1, with a "fresh trade" at their back and a hazy cloud cover veiling the morning sun, Green and Cook met on *Endeavour*'s quarterdeck—the far back region behind the ship's wheel and mizzenmast. With the swivel gun on its perch nearby, Green and Cook set up their sextants. Green's, built by the same Jesse Ramsden who'd created Banks's "electrical machine," was a fifteen-inch brass device on loan from the Royal Society. Cook owned his own—a twenty-inch, wooden-framed brass instrument with a pole that steadied it to the deck.[16]

Between 7:19 and 9:02 AM, the pair of explorer-astronomers made thirty-two angular measurements of the moon and sun's position. Green did most of the work. He found, for instance, the rising sun climbed from 21 degrees and 31 arc minutes to 45 degrees and 40 arc minutes over the 103-minute interval. The duo carried out multiple iterations of three repeated measurements of three values: the altitude of the moon's upper limb above the horizon, the altitude of the sun's lower limb and the angular separation between the nearest edges of the two bodies.[17] At the level of precision their calculations demanded, they even needed to keep track of whether their altitude measurements came from the *Endeavour's* quarterdeck (19–20 feet above the sea's surface) or the main deck or forecastle (18 feet). The *Nautical Almanac* then provided instructions and charts necessary to transform these sundry quantities into mariner's gold—longitude.

As Green found within thirty minutes of calculation, using inexpensive tools and methods that were coming into reach of practically any navigator on the planet, *Endeavour* lay at 14 degrees, 26 arc minutes north of the equator and 22 degrees, 47 arc minutes west of Greenwich. The world was becoming a smaller and more navigable place.

"In justice to Mr. Green," Cook later wrote, "I must say that he was indefatigable in making and calculating these observations." Cook also extolled Green's teaching skills. "Several of the petty officers can make and calculate [lunar longitudes] almost as well as himself," Cook continued. "It is only by this means that this method of finding the longitude at sea can be brought into universal practice—a method which we have found may be depended on to within a half a degree! Which is a degree of accuracy more than sufficient for all nautical purposes." The *Nautical Almanac* clearly had a convert.[18]

Two more days of northeasterly winds pushed the *Endeavour* steadily southward on its journey into a known dead zone in the Atlantic. Near the equator, the northeast trades die out, and ships enter a region called the doldrums—which for this mission started at about 12 degrees north latitude. Ships could spend weeks lolling around with the capricious currents and breezes. As Mark Twain would write more than a century later, the zone's prevailing characteristics are "variable winds, bursts of rain, intervals of calm, with chopping seas and a wobbly and drunken motion of the ship—a condition of things findable in other regions sometimes, but present in the doldrums always."[19]

Without any real progress to track on his sextant and *Nautical Almanac*, Green became a weather station for nine days, chronicling the passing clouds and "squally" storm fronts that knocked the boat about but rarely moved it much forward. Banks and Solander, by contrast, were overjoyed. Windless days for them meant the *Endeavour* became a kind of dedicated field biology lab. On October 4, Banks commanded a pinnace expedition into the calm seas to haul in and study various jellyfish and other many-tentacled creatures (like the "blue button" porpita) as well as water bugs and an exotic swimming mollusk called the blue sea slug (*Glaucus atlanticus*).[20]

Other days, like October 11, ended without taking a thing from the ocean but just observing its inhabitants from afar. "Saw a dolphin," Banks recorded. "And admired the infinite beauty of his colour as he swam in the water. But in vain. He would not give us even a chance of taking him."

On October 21, just north of the equator, the next trade wind picked up. The southeasterly trades that carry ships toward South America—the continent whose shoreline *Endeavour* would trace till it met the Pacific—required a different tack but constituted a reliable mode of transit nonetheless. The mission was back on course. Banks pleaded with Cook to detour to Fernando de Noronha, an archipelago of 21 islands 220 miles off the Brazilian coast. The captain agreed if the winds and current cooperated. They did not.[21]

Four days later, as a stiff morning wind fluttered flags and carried aloft all unmoored scraps of paper and cloth, Green stood on the same quarterdeck and used his same brass sextant to find the sun 77 degrees and 39 arc minutes above the horizon. According to his calculations, this put *Endeavour* at 0 degrees, 15 arc minutes south latitude. She had just crossed the equator.

"After we had got an observation and it was no longer doubted that we were to the southward of the Line, the ceremony on this occasion practiced by all nations was not omitted," Cook recorded in his captain's journal for the day. Anyone who couldn't prove he'd crossed the equator before, Cook added, had to make a choice: give up a bottle's worth of his rum ration or be ducked into the sea. Some twenty to thirty of Cook's men picked the latter option, Cook wrote, "to the no small diversion of the rest."[22]

Banks chose to pay out liquor rations for himself, his servants, and even his two greyhounds. Green does not record whether he chose dousing or not. His sober and matter-of-fact diary, however, suggests he doesn't seem the sort to have sorely missed four days' allotment of liquor. The cost, in Green's eyes, of a little rum seems slight, especially considering the ritual hazing of equator crossing he would have endured had he not paid it out.

Banks described the brutal hazing, a triple-ducking from the yardarm, which involved binding a sailor to a block of wood and hoisting him "as high as the cross piece [on the yardarm] over his head would allow." The dangling man would then be dropped into the ocean from the yardarm, the outermost tip of the horizontal spar that held the sails in place. "His own weight carried him down, [and] he was then immediately hoisted up again and three times served in this manner," Banks wrote in his journal. "Sufficiently diverting it certainly was to see the different faces that were made on this occasion, some grinning and exulting in their hardiness whilst others were almost suffocated and came up readily enough to have compounded after the first or second duck."[23]

RIO DE JANEIRO
November 13–December 7, 1768

To a veteran military leader like Antonio Rolim de Moura—viceroy of Brazil—the British cargo ship lumbering into Rio de Janeiro's harbor sent a message that anyone who knew espionage and privateering could understand. Count Rolim, as Cook would call the colonial governor, had seen too many dubious maneuvers by enemy Spanish forces to trust the English officers at their word.

According to representatives from the *Endeavour*, this English military mission (or was it a secret reconnaissance ship?) was stopping in Rio to restock so it could sail to some Pacific island and observe the planet Venus passing in front of the sun. And this so-called British Navy ship—one that stood apart from other British Navy ships on the oceans—carried onboard a retinue of "philosophers," equipped with advanced mapping and surveying technology, no less. Rolim no doubt thought his visitors were taking him for a fool. During the Seven Years' War and subsequent Portuguese-Spanish flare-ups in the New World, England had proved herself a conniving and untrustworthy ally. The prime minister of Portugal had sent instructions for Rolim's predecessor to remain vigilant against any British maneuvers in southern South America. The prime minister feared that England was working behind the scenes to take Brazil from Portugal just as it had taken France's colonial possessions in India.[24] *Endeavour*'s "philosophers" could just as easily have been undercover engineers on a mission to size up the Brazilian capital's defenses. Moreover, even if England had no designs on Brazil, King George III's ministers had financed smuggling missions that undermined the South American colony's economy. *Endeavour*, built to haul hundreds of tons of coal—or other booty such as Brazilian gold— was, after all, a smuggler's dream ship.

"The account we gave of ourselves," Cook later wrote in a letter to the Royal Society, "of being bound to the southward to observe the Transit of Venus (a phenomen[on] they had not the least idea of) appeared so very strange to these narrow minded Portuguese that they thought it only an invented story to cover some other design we must be upon."[25]

Lieutenant John Gore recorded in his diary that the viceroy viewed Banks and Solander as "supercargoes and engineers and not naturalists—

for the business of such being so very abstruse and unprofitable that they cannot believe gentlemen would come so far as Brazil on that account only."[26]

So the viceroy ordered *Endeavour* indefinitely detained, with only Cook and approved supply missions enjoying any access to the shore—and the captain was to be accompanied by a guard at all times.

"A boat fill'd with soldiers kept rowing about the ship," Cook wrote on the first full day in Rio, November 14. "Which had orders, as I afterwards understood, not to suffer any one of the officers or gentlemen except my self, to go out of the ship."

Cook petitioned Rolim to permit *Endeavour's* crew to make a supply run to shore. "But [Rolim] obliged me to employ a [Brazilian] person to buy them for me, under a pretense that it was the custom of the place. And he likewise insisted, notwithstanding all I could say to the contrary, on putting a soldier into the boats that brought anything to and from ship, alleging that it was the orders of his court and they were such as he could not dispense with. And this indignity I was obliged to submit to otherwise I could not have got the supply I wanted."[27]

Five days later, on November 19, crew and officials on *Endeavour* stifled their visceral reactions to the ordeal that Brazilian authorities had put one of *Endeavour's* boats through. Cook had sent his lieutenant Zachary Hicks ashore to learn why authorities continued to detain the ship and to plead again for an expedient end to the unnecessary delays. Cook told Hicks to return to *Endeavour* upon delivering the memo, but not to let any Brazilian soldiers onboard.

Rolim refused even to accept Cook's message. "Hicks . . . refused admitting a Portuguese sentinel into the pinnace," Green recorded in his journal. "Whereupon the boat's crew were drag'd to prison, Mr. Hicks made a prisoner and sent off to the ship quartered [on a Brazilian boat]. The pinnace [was] detained and his Britannic Majesty's Colours struck by the Portuguese."[28]

Banks, on the other hand, couldn't stand being cooped up while so much flora and fauna awaited cataloging onshore. Referring to the ancient

Greek myth of a king punished by the gods by keeping food and drink just out of his reach, Banks wrote to the Royal Society, "All that we so ardently wish'd to examine was in our sight. We could almost but not quite touch [the onshore specimens]. Never before had I an adequate idea of Tantalus's punishment."[29]

Knowing the penalties if caught, Banks nevertheless snuck ashore once under cover of night—and sent his servants ashore at least two other times as well. "I myself went ashore this morn before day break and stay'd till dark night," Banks wrote on November 26. "While I was ashore I met several of the inhabitants who were very civil to me, taking me to their houses where I bought of them stock for the ship tolerably cheap: A porker middlingly fat for 11 shillings, a muscovy duck something under two shillings, etc. The country, where I saw it, abounded with vast variety of plants and animals, mostly such as have not been described by our naturalists as so few have had an opportunity of coming here."[30]

Parkinson, Banks's young artist, described one late-night foray into the Brazilian beyond. "Having obtained a sufficient knowledge of the river and harbour by the surveys we had made of the country," Parkinson wrote in his journal, "we frequently, unknown to the [Portuguese] sentinel, stole out of the cabin window at midnight, letting ourselves down into a boat by a rope; and, driving away with the tide until we were out of hearing, we then rowed to some unfrequented part of the shore, where we landed and made excursions up into the country."[31]

However, others snuck ashore for less scientific study. One American-born midshipman on *Endeavour* left behind a journal account of an undercover trip into Rio where he discovered that the city's "genteeler prostitutes . . . make their assignations at church."[32]

And so through the end of November, Cook continued with what he called his "paper war between me and his Excellency wherein I had no other advantage than the racking of his invention to find reasons for treating us in this manner for he never would relax the least from any one point." Cook and his lieutenants stood flummoxed by the intransigence of the viceroy and his court. On the other hand, the viceroy's concerns weren't imaginary either. Just five years earlier, Brazil had relocated its capital from the booming city of Bahia to the smaller but more strategically situated Rio.[33] If *Endeavour* had been on a mission to smuggle or reconnoiter Portuguese defenses, Rolim could scarcely have said he didn't see it coming. The English boat and its proclaimed mission were conspicuously strange. It didn't help matters that Cook's pride sometimes appeared to camouflage what Rolim saw as dubious motives. Cook had told the viceroy that his journey from Madeira was brief and uneventful. And Cook had admitted that at Madeira he'd stocked the ship to the gills. So, Rolim said, "Why did you want so soon water and provisions? It could only proceed from not having loaded a sufficiency of those articles in that island."[34]

As a Spanish proverb from the time put it, "I can take care of my enemies, but God protect me from my friends."[35]

So England and Portugal remained symbolically locked in the friendly but adversarial embrace of the former's ship in the latter's harbor. Ultimately, however, Rolim could find nothing more damning than the unusual nature of *Endeavour*'s mission. He saw no overriding reason to justify holding Cook and his ship beyond December 2, when Rolim gave his blessing for *Endeavour* to proceed on its way.

Not coincidentally, Cook made his most extensive diary entry to date soon thereafter, on December 7, when the captain chronicled "the Bay or River of Rio de Janeira." Rolim's (justifiably) paranoid insistence to the contrary, Cook had no mission to reconnoiter Rio. But as if to prove that he could do it anyway, Cook took pains in his journal to describe the kind of military details the commander of an invading force might want to know about the city and its defenses.

"I shall now give the best description I can of the different forts that are erected for the defense of this bay," Cook wrote. "The first you meet with coming in from the sea is a battery of 22 guns . . . to hinder an enemy from landing in that valley . . . whence I suppose they may march up to the town."[36]

TIERRA DEL FUEGO
January 6–16, 1769

Condensed breath warmed fingertips exposed to the frigid South Atlantic gales. Captain Cook had issued all men onboard *Endeavour* a thick jacket named after the first European explorer who'd braved the far southern latitudes en route to the Pacific. "All hands bend their Magellan Jackets (made of a thick wollen stuff . . . call'd fearnought)," Banks recorded in his diary on January 6. "And myself put on flannel jacket and waistcoat and thick trousers."

One of Cook's two lieutenants, John Gore—as well as the ship's master Robert Molyneux and two of his master's mates—had sailed along these same straits just two years before. Their experiences at the southern tip of South America had already entered English naval lore.

The HMS *Dolphin* and HMS *Swallow* had spent four months weaving through the maze of inlets and bays constituting the Strait of Magellan. And when they emerged into the first open stretch of the Pacific Ocean in April 1767, a strong Pacific current separated the boats, permanently. So a joint expedition consisting of two British naval vessels involuntarily split into two separate voyages.

The *Endeavour* could not spare four months. Even four weeks spent turning this corner of South America from one ocean to another, and the mission might miss its entire purpose—perhaps still wandering the Pacific as Venus made its brief voyage across the sun's disk.

On the night of January 6, Joseph Banks held fast to his swinging bed, its brass holdings creaking with each sway. Banks and the mission's other supernumeraries, encased in webbed rope beds hung from the rafters, followed the ship's each sway and jolt.[37] "The evening blew strong," Banks recorded, shivers running through the spines of his words. "At night a hard gale, ship brought to under a mainsail; during the course of this my bureau was overset and most of the books were about the cabin floor, so that with the noise of the ship working, the books etc. running about, and the strokes our cots or swinging beds gave against the top and sides of the cabin, we spent a very disagreeable night."[38]

These were days that tested every man. The captain, as taxed by the elements as any of his crew, had opted for this stormy route. *Endeavour* could have just struck out to sea till it hit 61 or 62 degrees south latitude—a parallel that lay safely beyond the hidden shoals and mast-splintering tempests that made the far end of South America infamous. The ocean to the west, in fact, took its name from the pacific calm it welcomed all sailors with who had survived the Cape Horn crossing.

However, Cook later wrote, "As to running into the latitude of 61 or 62 degrees before any endeavour is made to get to the westward, it is what I think no man will ever do who can avoid it, because it is not southing but westing that is wanted. This way, however, he cannot steer, because the winds blow almost constantly from that quarter [i.e., the west]."[39]

Moreover, the coast of Tierra del Fuego offered supplies, giving the captain yet one more reason to hug the dangerous coastline. It was no Garden of Eden like Madeira, but the extreme tip of South America would be the last landfall *Endeavour* might reasonably expect to make before she arrived at Tahiti. Who knew how many months thereafter the ship would be living off the rations in its stockrooms?

On January 11, *Endeavour* sailed into view of Tierra del Fuego—an island separated from the South American mainland by the Straits of Magellan. The calm weather welcomed the sub-Antarctic island's guests. "We could see trees distinctly through our glasses and observe several smokes made probably by the natives as a signal to us," Banks wrote in his diary. "The captain now resolved to put in here if he can find a convenient harbour and give us an opportunity of searching a country so entirely new."[40]

Four days later, after tacking back and forth between Tierra and the nearby Staten Island, the winds and weather permitted a resting place in calm waters. *Endeavour* anchored in the Bay of Good Success, an inlet on the eastern end of the Cape Horn gauntlet. Banks and Solander rowed ashore after the noon meal and, armed with trifles, greeted the natives. "Dr. Solander and myself then walked forward 100 yards before the rest and two of the Indians advanc'd also and set themselves down about 50 yards from their companions. . . . We distributed among them a number of beads and ribbands which we had brought ashore for that purpose, at which they seem'd mightily pleased."[41]

Cook noticed the locals' familiarity with the Europeans' guns and dyed goods. "They were not at all surprised at our fire arms," Cook wrote. "On the contrary seem'd to know the use of them by making signs to us to fire at the seals or birds that might come in the way. . . . They are extremely fond of any red thing and seemed to set more value on beads than any thing we could give them. In this consists their whole pride; few either men or women are without a necklace or string of beads made of small shells or bones about their necks."[42]

Finding as welcoming a rest stop as could be hoped, Cook gave Banks permission to venture farther inland to collect specimens while *Endeavour*'s crew restocked the ship. The next morning, January 16, a sunny summer day in the sub-Antarctic, the ship's three scientists plus a support crew set out to explore Tierra del Fuego.

Banks, Solander, Green, two assistants, four servants, two seamen, and a midshipman pushed inland from the beach into a thick grove of trees, on an ascending path. "Neither heat nor cold was troublesome to us nor were there any insects to molest us," Banks recorded.[43]

Once they'd reached the top of their summit, however, only mire awaited. Low birch bushes cut at their sides while their boots sank into muck. "Every step the leg must be lifted over [the bushes] and on being plac'd again on the ground was almost sure to sink above the ankles in bog," Banks noted. A rocky outcropping seemed close at hand, though, so the party pressed on.

Then, during the unexpectedly strenuous walk to their new destination, Banks's hired sketch artist Alexander Buchan had an epileptic seizure. "A fire was immediately lit for him and with him all those who were most tir'd remained behind, while Dr. Solander, Mr. Green, Mr. Monkhouse and myself advanced," Banks wrote.

However, snow squalls—which were not uncommon on this island of minute-to-minute weather variability—descended on the group. Facing whiteout conditions and plummeting morale, Banks changed his tack. He realized that with one man in an uncertain state of health and others beginning to succumb to cold, someone had to take charge. Gratefully, Banks had untapped leadership skills in him—no doubt augmented by learning a few things from his extraordinary ship's captain. The gentleman naturalist now began delegating people to start a real fire and prepare a shelter for the night. They were going to have to ride out the storm.

"The air was here very cold and we had frequent snow blasts," Banks recorded. He sent Green and Monkhouse back to the group that was tending to Buchan. They'd all rendezvous at a nearby hill, Banks decided, and fashion a wigwam out of trees and branches for the night. But now the first signs of hypothermia began to appear. "We pass'd about half way very well when the cold seem'd to have once an effect infinitely beyond what I have ever experienced. Dr. Solander was the first who felt it. He said he could not go any farther but must lay down, tho the ground was covered with snow. And down he laid notwithstanding all I could say to the contrary."

One of the servants, who carried a jug of rum, began drinking to keep warm. He started growing tired even faster than the rest of the group.

"Richmond, a black servant, now began to lag and was much in the same way as the doctor," Banks noted. "With much persuasion and entreaty we got through much the largest part of the birch [thicket] when they both gave out. Richmond said that he could not go any further, and, when told that if he did not he must be froze to death, only answer'd that there he would lay and die."

A distant planet crawling across a fiery star's face might have warmed the thoughts of *Endeavour*'s natural philosophers. However, in a few hours apart from the ship, the entire mission had shifted from cutting-edge science to simple survival. Banks roused Solander to the birch shelter and had sent two others to try to retrieve Richmond. Only one of Richmond's rescue party, frostbitten and delirious, made it back to the lean-to. "The road was so bad and the night so dark that we could scarcely ourselves get on nor did we without many falls," Banks wrote. "Peter Briscoe, another servant of mine, now began to complain and before we came to the fire became very ill but got there at last almost dead with cold."

"Now might our situation truly be called terrible," Banks continued. "Of twelve of our original number were 2 already past all hopes, one more was so ill that tho he was with us I had little hopes of his being able to walk in the morning. . . . Provision we had none but one vulture which had been shot while we were out. . . . And to compleat our misfortunes we were caught in a snow storm in a climate we were utterly unacquainted with but which we had reason to believe was as inhospitable as any in the world."

Chapter 9

A SHINING BAND

VARDØ, NORWAY

January 10–May 9, 1769

In a world of endless night, there's never enough time for sleep. Father Maximilian Hell and his scientific assistant, Joannes Sajnovics, had discovered this counterintuitive fact of life in their arctic barracks as their "days" and "nights" bled into a hazy purgatory of perpetual sunlessness. "The winds are always raging," Sajnovics later wrote to his Hungarian Father Superior. "You would think these people have plenty of time to sleep. But on the contrary, we have never slept so little as we do on these long nights."

Without a solar cue to demarcate public and private social hours, Sajnovics found visitors streaming in to the voyagers' makeshift house at increasingly impolite times. "People visit one another during the night here and offer their guests coffee and tea, just like during the day," Sajnovics wrote. "We also have chocolate, but that is disappearing very fast."[1]

On January 10, the voyagers and soldiers completed the island's new observatory, a long, single-story wooden shed with sliding hatches in the roof and walls to enable observations of the skies when weather cooperated. The astronomers had much to do in the months before the transit. Observations of stars' maximum altitudes could, when compared to the same measurements at a known latitude, yield the observatory's latitude. That was the easier of the observatory's two terrestrial coordinates. Hell and Sajnovics had hoped to discover their longitude by timing an eclipse of the moon two days before Christmas. But, as Sajnovics later wrote in a letter from Vardø, "The famous lunar eclipse from 23 December passed invisibly above us to our great sadness." In his diary entry for the day, Hell's assistant noted, "We could not see any trace of the moon because of the curtain of clouds. We responded to the untactful mockery of the [village's] priest with a few stinging replies."[2]

Instead, they tried to find their longitude via observations of Jupiter's moons. But the planet was so close to the horizon that it was difficult to resolve in their scopes. Lacking marine chronometers, eclipses, or Jovian lunar data, they had to use the moon. But despite being editor of a leading lunar almanac, Hell didn't put much stock in the lunar longitude method as anything more than an approximation.[3] So Vardø remained, for the time being at least, effectively longitudeless.

Other careful measurements demanded the scientists' attention. First, the pull of gravity was slightly stronger in Vardø than it was in central Europe. This meant a second as measured on their pendulum clock would be slightly shorter than the time interval the same pendulum clock would tick out in Vienna. Figuring out exactly how much shorter involved timing individual stars' progress through the sky for a complete twenty-four-hour rotation of the earth. Hell and Sajnovics then used this data to fine-tune the length of their clock's pendulum so that it ticked out something much closer to true seconds, minutes, and hours. Furthermore, Hell suspected the arctic atmosphere was thicker, which would increase the amount by which the sun's light is deflected (or refracted). If so, another set of corrections would need to be applied to account for Vardø's different levels of atmospheric refraction.[4]

Meanwhile, each day Sajnovics described the living conditions on the iced-over chunk of rock that was now their home. "To the west the island has a about a quarter mile of land; the rest is covered with a thick iced glacial sea covering about one mile," he wrote in a letter from Vardø. "And even though it is flat, it is here and there pierced by coarse layers of rocks. This environment does not produce anything besides moss and always available cochelaria [scurvy grass]. . . . There are no seedlings or trees on the island, nor on the neighboring lands. They transport the firewood from many miles away, mainly from Russia."[5]

The winter was, as the voyagers had expected, brutal. "We were waiting for better days with peace and patience and we were expecting the sunshine to disperse the darkness," Sajnovics wrote in another letter. "We were sitting at home making proper use of our time, reading books, defining sea algae and examining snails, hoping that one day people might benefit from the fruit of our loneliness."[6] But even after the sun began to reintroduce itself in mid-January, the storms and long nights still took a toll. "Nothing special," Sajnovics bleakly journaled on March 8, "unless the great coldness and the wind are worth mentioning."

Hell had passed part of the time plying his spiritual trade, counseling some of the local residents in their no doubt seasonally affected malaise. According to Sajnovics, Hell helped to settle some family disputes and personal grudges among the populace—to the point that some formerly feuding parties now amicably visited with one another and even held gatherings together.

"These good people thought we were very much like them, and we would spend the whole night playing cards and going to dances with them," Sajnovics wrote in a letter from Vardø. "So they immediately decided on a day when they would hold a night party at the commander's place every week. . . . But soon they noticed from the way we talked and behaved that the Roman Catholic priest has a certain dignity—admiring us for our lonely lifestyle, moderation, soberness, and especially for the way we were looking away from the ladies."

Hell and Sajnovics's newfound affection for the locals was partly due to a surprise discovery Sajnovics had made: The regional Lapland dialect, Sami, was a linguistic cousin to Hungarian.[7] Hell was overjoyed at the discovery of his newfound cousins. "They are Hungarians," Hell wrote in a letter to a Hungarian Jesuit colleague, thanking the creator of the universe he hoped to unlock. "They speak our language; they wear our Hungarian clothing, they live according to the customs of our Hungarian forefathers. In a nutshell, they are our brothers!"[8]

VARDØ, NORWAY
May 27–June 6, 1769

One week before the century's final Venus transit, on Saturday, May 27, nature put on a show for the visiting scientists in Vardø. The meteorological and northern lights spectacle was either a sign of good grace toward the Jesuit astronomers or a faith-testing demonstration of how badly things could go wrong in seven days. Weather along the north Norwegian Sea coast during June is often foggy, from nonstop sunlight evaporating arctic waters and offshore breezes blowing the moist air inland.[9] Hell and Sajnovics knew they had a fair chance of never seeing the Venus transit. But something told them God was smiling on the voyage—although that didn't stop the voyagers from bettering their odds by praying often and, as Sajnovics records in his journals and letters, taking Communion not infrequently too.

Whatever the source, the snow clouds flirting with sun and "polar lights" on May 27 did provide the visitors wonders to marvel at. Vardø's mayor visited the observatory in the morning, as the thirteenth straight day without any night wore on. "We saw a remarkably beautiful northern light east to the sun," Sajnovics wrote in his journal. "The sun was standing high above the horizon between north and west covered by a dense snow cloud. Thirty degrees to the northeast there was another dense snow cloud at the same altitude. The rest of the sky was clear.

Very beautiful rays of light were stretching from the former cloud into the latter in great number; they were stretching long from the northwest and all the way to the zenith with bright particles rolling with an incredible speed towards the cloud that was in the northeast."

At noon, the three astronomers—Hell, Sajnovics, and Borchgrevink—manned the Vardø observatory's two quadrants. They used the devices to find the sun's highest altitude, at noon. The chief instrument for the task, loaned to the expedition by German explorer Carsten Niebuhr, was a "portable" piece of bronze and iron optical machinery that had been carted around in three big boxes. According to Hell's manuscript account of Niebuhr's quadrant, a trained technician could set the instrument up or take it down in an hour. Its three heavy iron feet formed the base on which the brass optical finery rested. The quadrant looked a little like a quarter slice of pie made of trelliswork. A small telescope was hinged at the pie slice's center with its eyepiece fixed along the circular edge by brass rollers attached to a brass plate. Tick marks and subdivisions (along a "Nonius" or Vernier scale) along the circular segment enabled angular measurements of the telescope's orientation down to its finest scale, 30 arc seconds. The quadrant had its own plumb line, which gave them their true vertical and horizontal directions.[10] A trained user of these astronomical quadrants could in a few seconds' time get a precise fix on a sun, star, or planet's altitude measured in degrees, arc minutes, and tens of arc seconds.[11]

Niebuhr had famously used the same quadrant to become the first European to map the Arabian peninsula during a disastrous trip to the Middle East and India from which he was the only survivor. Hell sometimes complained about Niebuhr's quadrant, but his team's measurements using the instrument were nevertheless well beyond merely adequate.[12]

A storm then dumped snow on the island and enveloped the sun. Although it kept the region in daylight twenty-four hours a day, the sun remained hidden behind the clouds for the ensuing week. A quiet six days of tense preparations passed before Sajnovics even dared to offer up his mission's fate to the powers above. "We did the necessary preparations for the observations tomorrow," he wrote on June 2. "If it pleases God—Will's Gott! [Let it be God's will!]"

By 3:00 AM the next morning, the sun had shone briefly and just as quickly disappeared behind a curtain of clouds. At 9:00 AM, some twelve hours before the transit, the astronomers held mass. Tenuous and wispy streams of cirrocumulus—or perhaps northern lights—danced across a windy sky. Tension mounted throughout the afternoon, as the sun peeked through the sky's cottony blanket and then hid beneath it again like a playful child. Sajnovics, a gourmand who often took pains to discuss food and drink throughout his travel diaries, didn't bother to describe having even a drop of water throughout a very long June 3 Venus transit day. As the evening progressed toward the fated nine o'clock hour, the mood in the observatory began to darken.

"We had put all our trust in God," Sajnovics later wrote in a letter to his Hungarian Father Superior. "At nine in the evening, we were standing in front of the telescope, torn between fear and hope—Hell and me and the student [Borchgrevink], trying to see the entrance of Venus into the rising sun. [But] lo and behold, there was an opening in the clouds and we could see the sun as if through a window!"[13]

According to Hell's published account (in Latin) of the Venus transit at Vardø, at 9:15 PM and 17 seconds (9:15:17), the first hints of Venus's leading edge began encroaching into the sun's disk.[14] (The observers later corrected these observations for effects like atmospheric refraction. The "true" first ingress time then became 9:16:39.8. Hell's group's claims of tenths-of-a-second accuracy is laudable but doubtless, removed from the excitement of the moment, errs on the side of ambitiousness.)[15] Over the ensuing quarter hour Venus's silhouette grew in the sun from a tiny dot to a little dimple to a dark semicircle to a dark circle with just a little wedge taken out. Then, at 9:32:30, Sajnovics was the first of the three to yell out that he could see the whole of Venus's circumference against the sun. At that moment, the sun appeared to the men as a giant luminous globe out of which a black circle 1/35 of the sun's apparent diameter had been stamped out. For six seconds starting at 9:34:04.6, the very temporary alignment of these celestial bodies enabled the sun to directly illuminate Venus's atmosphere and create a kind of brilliant thread that traced Venus's circumference. Hell described the momentary spectacle surrounding the tiny planet as a "shining band."[16]

The shouting and commotion inside the observatory kept no secrets about the stunning moment of triumph happening within its walls. The Vardø town storekeeper, overjoyed at his visitors' great success, took a swig of liquor and shot off his big guns. "The merchant fired his 9 cannons and raised the flag in order to express his joy," Sajnovics wrote. "The commander followed his example and put out the flag of the fortification. We allowed people into the observatory and showed them the planet Venus inside the sun. But they could only see it for 5 minutes, because the sun was then covered by dark clouds."

The sun had returned to obscurity. The observers' job was only halfway done. Without recording similarly precise exit times—the exact moment when Venus's outer edge reaches the sun's inner edge and when the last hints of Venus's silhouette completely disappear from the solar disk—halfway done was almost as good as nothing at all. "It was cloudy for five hours in a row," Sajnovics later wrote in a letter. "So we had to give up on the prospect of seeing [Venus's egress]. The guests were standing sad beside us with sour expressions—and shared their sympathy and pain in silence. You cannot imagine what we felt."

Yet mere minutes before Venus was to conclude its solar passage, the wall of clouds again parted to clear a line of sight. "We were amazed in our great joy and praised the Heavens," Sajnovics later wrote. At 3:26 AM and 6 seconds, Hell saw the first hint of the optical illusion that just eight years before had made precision timing of the transit so unexpectedly difficult. The black drop effect (as it was later termed) may have all but ruined most of the 1761 Venus transit observations, but Hell and Sajnovics were prepared for it. Four seconds later, Borchgrevink shouted out his sighting of Venus's exiting internal contact. Sajnovics followed suit at 3:26:18. As with the planet's ingress earlier in the evening, a quarter hour passed and the trio trained their eyes on their century's final glimpse of a transiting Venus. Borchgrevink was first to call out his judgment that the celestial phenomenon was officially over, at 3:44:20. Four seconds later, Hell marked down that the Venus transit—the greatest celestial event of his lifetime—was finally and certainly complete. Sajnovics signed out a second after that, at 3:44:27.[17]

And so, despite impossible observing conditions, Hell and Sajnovics's Venus transit expedition enjoyed every success it could have hoped for.

"The merchant fired his three howitzers again six times," Sajnovics wrote the next morning. "Everyone in Vardø was very very happy. We said a Te Deum Laudamus . . . with all our heart, then we retired to bed." To a pair of Jesuit astronomers and adventurers, this Catholic hymn of praise offered up to their maker every thanks a human soul could offer. "It is sung in the Romish church," explained one contemporary encyclopedia, "with great pomp and solemnity upon the gaining of a victory or other happy event."[18]

Sajnovics wrote in a letter a few days later, "Our guests find the word 'miracle' unusual, perhaps even ridiculous. Yet they all agreed that it was not merely due to regular occurrence of natural phenomena, but it was due to some special divine intervention, that we could see the astronomical phenomenon with such clarity and in such terrible weather, as if through magic, in the most clear sky possible. I will remember this miracle for as long as I live."[19]

Chapter 10

FORT
VENUS

Southern Pacific Ocean
January 17–March 26, 1769

Charles Green, Joseph Banks, and Daniel Solander survived the ordeal at Tierra del Fuego, although two of their servants did not. The weather had broken the following morning, and after a frigid three-hour march the team made it back to the beach. "On reviewing our track as well as we could from the ship, we found that we had made a half-circle round the hills instead of penetrating as we thought we had done into the inner part of the country," Banks recorded on the day he feared he might not live to see. "With what pleasure then did we congratulate each other on our safety no one can tell who has not been in such circumstances."[1]

Having rounded Tierra del Fuego in stormy and foggy weather, Cook ordered *Endeavour* on a southwesterly course to ensure that she was clear of Cape Horn and its unwelcome surprises. For three days and nights at the end of January, Green supervised seventy-eight observations of sun and moon, cracking the spine on his new *Nautical Almanac* for 1769. On January 30, Green, midshipman Jonathan Monkhouse, and Banks's scientific secretary Herman Spöring set up the ship's sextants on *Endeavour's* main deck, seizing moments of clear throughout the squally day. The three measured—in line with *Nautical Almanac* instructions—angular distances between sun and moon and altitudes of the two bodies from the horizon. Green found that *Endeavour* was 60 degrees and 2 arc minutes south latitude, 73 degrees, 27 arc minutes, and 50 arc seconds west of Greenwich.[2] Cook was satisfied that all traces of South America lay in the ship's wake.

Now a 4,000-mile path across a new ocean stretched out ahead.

The Pacific Ocean was as mysterious an entity as could be found anywhere on earth. Most of it was uncharted, and indeed some eminent scholars claimed that a great undiscovered continent—something so vast as to offset the mass of Europe, Asia, and North America—waited somewhere in the Pacific's southern reaches.

More immediately, winds and currents remained as unknown as the landmasses that might spring up along the journey. A broad ocean upwelling that would later be called the Humboldt Current—named after a scientist born, as it happens, in 1769—brought much of the southern Pacific's aquatic life to the *Endeavour* as she cut the first part of her course to the northwest. "The weather was such," Cook recorded in early February, "as to admit Mr. Banks to row round the ship in a lighterman's skiff shooting birds."[3]

Banks's ornithological specimens grew by the bushel, with a few hearty meals of bird meat tucked in as well. Banks kept his sketch artists fully employed too, portraying both the tropical birds and fish being hauled in as well as the still unrecorded specimens from earlier in *Endeavour's* voyage. "This morn some sea weed floated past the ship and my servant declares that he saw a large beetle fly over her," Banks recorded on February 9. "I do not believe he would deceive me, and he certainly knows what a beetle is, as he has these 3 years been often employ'd in taking them for me."[4]

And so life onboard a tiny, three-decked ship with more than ninety men carried on through February and into March. Some days produced sun; other days produced rain. Each day marked tens or even scores of miles' progress toward the mission's Tahitian destination. An uninitiated visitor descending on *Endeavour* in full sail through the South Pacific might—after first marveling at its astonishingly cramped quarters— have to stifle a gag reflex: the warmer weather meant increasingly spoiled meats and stocks. *Endeavour*'s freshwater supplies were becoming breeding pools of algae and slime. The pigs, poultry, and goats added their own noxious perfume to the moist air. Allowing the sweat-drenched crew to bathe did little to abate the problem. The Royal Navy was still more than a decade away from introducing soap to its hygienic arsenal. Special sails Cook rigged to help vent the putrid gases below decks gave little relief. And no amount of romanticizing this indisputably historic voyage can gloss over the maggot- and cockroach-laced foods—often gnawed on by rats too—that the men had to down. Wiser crew members waited till nightfall to eat. At least that way they didn't have to see their meal.[5]

Liquor provided what little gastronomic relief might be enjoyed. The beer was kept somewhat palatable by mixing in flour, sugar, and salt. Add rum or brandy and the concoction was called "flip." "Grog" was water plus rum. A day's ration of alcohol amounted to about five shots (250 ml), which filled out the day's liter of water giving a man a slight buzz without inducing drunkenness.

The ship's bow held what the crew called *Endeavour*'s two "seats of ease," the toilets. All parts exposed on these seats—dropping straight into the ocean—felt the brunt of the oncoming wind and surf. And if a wave crashed while one sat doing one's business, so be it. Welcome to His Majesty's Navy.

One Saturday in late March, a marine marched up toward the seats of ease. Any who saw him make his way thought nothing of the commonplace routine. But it was dusk, and when the marine hadn't returned from the dark a half hour later, his fellow men at arms began to wonder.

"At 7, William Greenslade, Marine, either by accident or design, went overboard and was drowned," Cook laconically noted in his diary. Banks's journals reveal Greenslade headed toward the bow that evening never intending to return.

"He was a very young man, scarce 21 years of age," Banks noted in his diary. "Remarkably quiet and industrious."

Greenslade had been guarding the cabin door when one of Cook's servants pulled out a piece of sealskin for making tobacco pouches. Greenslade asked for a piece of the skin but was refused. The marine surreptitiously helped himself to the sealskin nevertheless, but when the servant discovered the theft, he grabbed it right back. And so the matter would have ended there, with the servant saying he wouldn't snitch since ultimately there was no crime. But the sergeant got wind of the offense and put it into his head that the honor of the marines was suddenly at stake. The Articles of War, read aloud to the entire ship's crew every Sunday, made stealing aboard a naval vessel a capital crime. Replacing executed crew members at sea was a real difficulty, though, so Greenslade would probably have been whipped or otherwise humiliated in front of the rest of the ship. Clearly, however, to this soldier, death was preferable to dishonor.

"To make [Greenslade's] exit the more melancholy," Banks added, "[he] was drove to the rash resolution by an accident so trifling that it must appear incredible to every body who is not well acquainted with the powerful effects that shame can work upon young minds."[6]

THE SOUTH PACIFIC AND TAHITI
April 4–17, 1769

By early April, *Endeavour* had spent nearly two months crossing open ocean without a hint of land, let alone any fabled lost continent. And then on April 4 at 10:30 in the morning, Cook finally spied his first sight of terra firma, a tiny oval-shaped atoll just six miles across at its widest. "I named it Lagoon Island," Cook noted in his diary entry for the day.[7] "We approach'd the north side of this island within a mile and found no bottom with a 130 fathom of line, nor did there appear to be any anchorage about it. We saw several of the inhabitants, most of them men. And these march'd along the shore abreast of the ship with long clubs in their hands as tho they meant to oppose our landing."

Endeavour sailed on. The next day she encountered another coral atoll, this one uninhabited. Cook named it Thrum Cap, with the following day's land-ho (again uninhabited) being dubbed Bow Island. So *Endeavour* atoll-hopped for the following six days until on April 12 its destination came into view. Green, Cook, and two of *Endeavour*'s officers stationed themselves on the main deck, sextants in hand, to perform the familiar task of taking the three essential measurements of moon and sun (respective altitudes and angular separation) and repeating the measurements each three times.[8] Another half hour's scribbling, and the *Nautical Almanac* yielded their longitude approaching Tahiti—148 degrees and 58 arc minutes. Cook estimated at the time the island lay to the southwest "distant 6 or 7 leagues," which would make the calculated longitude something close to exact.

Endeavour's captain could—and did—take pride in a great medical accomplishment. In addition to the previously noted drowning, suicide, and freezing deaths on Tierra del Fuego, Cook had lost only one more crew member (another drowning) during the seven and a half months *Endeavour* had been at sea. Ships traveling the distances *Endeavour* plied often crewed double the hands they needed, expecting to lose scores of men to scurvy. Cook hadn't lost a single soul to the disease.

On a sunny Thursday morning, April 13, as *Endeavour* ran an easy sail into the bay that would be its anchorage at Tahiti, Cook took stock of his record. In his estimation, he wrote, the rendered meat products and sauerkraut constituted the main reason for his defeat of the dreaded seaman's plague. The captain allowed himself to gloat a little. Every day, he said, he'd ordered the sauerkraut "dressed . . . for the [officers'] cabin table, and permitted all the officers without exception to make use of it—and left it to the option of the men either to take as much as they pleased or none at all." When the officers, by Cook's orders, were seen taking such pleasure in eating the foodstuff, seamen began clamoring for it. Not even a week later, Cook began rationing sauerkraut because it had become so popular.

"Such are the tempers and disposition of seamen," Cook noted. "Whatever you give them out of the common way . . . you will hear nothing but murmuring 'gainst the man that first invented it. But the moment they see their superiors set a value upon it, it becomes the finest stuff in the world."[9]

As with Cook's experience at Lagoon Island, the native inhabitants of Tahiti beat a hasty path to the beach once word had spread that *Endeavour* lay anchored offshore.

"This morn early came to an anchor in Port Royal bay, King George the Third's Island [Tahiti]," Banks wrote in his journal. "Before the anchor was down we were surrounded by a large number of canoes who traded very quietly and civilly, for beads chiefly, in exchange for which they gave cocoa nuts, bread-fruit—both roasted and raw—some small fish and apples."

The Tahitians also tried to sell a pig to *Endeavour*'s crew, who offered up some nails in exchange. The islanders were already, because of their previous encounters with Europeans, savvy hagglers. The Tahitians, Banks wrote, "repeatedly offer'd it for a hatchet; of these we had very few on board, so thought it better to let the pig go away than to give one of them in exchange, knowing from the authority of those who had been here before that if we once did it they would never lower their price."

The Tahitians' open society was already well-known. Before the day was out, Cook had laid down five rules for his crew to govern their interactions with the locals. The first mandated the men to "cultivate a friendship with the natives and to treat them with all imaginable humanity." Not two years before, the HMS *Dolphin* had committed a minor massacre on the natives after a pathetic volley of stones thrown from canoes bumped off the ship's hull. Cook wanted no repeats of that mistake.

The second, third, and fourth of Cook's rules governed trade with the locals, limiting such dealings to a delegated representative and stipulating penalties for trading or losing their own or the ship's goods. The fifth rule, though, was a preventative measure that spoke to social peculiarities specific to Tahiti.

"Fifth," Cook spelled out, "No Sort of iron, or any thing that is made of iron, or any sort of cloth or other useful or necessary articles are to be given in exchange for any thing but provisions."[10]

Tahitians, as the crew on the *Dolphin* had discovered, loved iron. They also lived in a fecund society that imposed precious few restrictions on sexual intercourse—and boasted naturally large extended families as a result. From their early teenage years onward, Tahitian boys and girls were encouraged to unloose their libido as often as they liked—so long as the copulating couple was not known to be directly related and did not cross class lines.[11] For HMS *Dolphin* sailors who could sneak off to the island, iron quickly became the currency of sex. Before *Dolphin*'s five-week Tahitian stay was over, a seaman could hardly find a loose nail or hook to hang his hammock on. Here was another line that Cook did not want *Endeavour*'s men to cross.[12]

Endeavour's landfall at Tahiti was marked, Cook recorded, by a mutually edgy and mistrustful peace. The morning of Friday, April 14, Cook, Banks, Solander, and a few of the ship's officers made a foray inland to find the best location for their ship (offshore) and observatory (onshore). Hundreds of locals greeted *Endeavour*'s noteworthies. "Mats were spread and we were desired to sit down fronting an old man who we had not before seen," Banks recorded. "He immediately ordered a cock and hen to be brought which were presented to Captain Cook and me. We accepted of the present."

Banks's wandering eye lighted on a young beauty in the crowd. The Englishman was seated with the chief's wife, though, who Banks noted was "ugly enough in conscience." Banks began trying to impress the nubile young lady, "load[ing] my pretty girl with beads and every present I could think pleasing to her," Banks wrote. But the courtship ended when Solander and William Monkhouse, the ship's surgeon (brother to *Endeavour*'s midshipman), discovered that their pockets had been picked.

"Complaint was made the chief," Banks added. "And to give it weight I started up from the ground and striking the butt of my gun made a rattling noise which I had before used in our walk to frighten the people and keep them at a distance." After some delicate negotiations, the missing opera glasses and snuff box were returned—although the snuff itself had gone missing.

The following day was worse, with a detachment going ashore and ultimately shooting dead a Tahitian who tried to pilfer a musket. Buchan, the painter who had suffered an epileptic fit at Tierra del Fuego, had a second seizure on April 16 and died the next day.

"His loss to me is irretrievable," Banks recorded. "My airy dreams of entertaining my friends in England with the scenes that I am to see here are vanish'd." Out of fear of what the Tahitians might do with Buchan's body, Cook ordered a small crew to pilot one of *Endeavour*'s boats out to open water, where the artist was given a sea burial.

Ashore, the Tahitians had attempted a peace offering of some breadfruit and a few hogs. Cook gave the two chiefs who'd brought forward the offering a nail and a hatchet each. Cook had begun sizing up the shore for his ultimate purpose of visiting Tahiti—an astronomical event now forty-seven days away.

"In the afternoon," Banks recorded, "we all went ashore to measure out the ground for the tents, which done, Cap Cooke and Mr. Green slept ashore in a tent erected for that purpose, after having observ'd an eclipse of one of the satellites of Jupiter."[13]

TAHITI

May 1–2, 1769

Theft had become practically a form of communication between Cook's men and the Tahitians. European explorers visiting the island—who claimed the whole landmass first for Spain, then for France, and then for England—usurped big. But Tahitians often picked pockets and pilfered whatever they could lay their hands on whenever the Europeans came into their presence.

Green, Banks, Solander, and Cook had spent the remainder of April leading *Endeavour's* carpenters and crew in building Fort Venus—an emerging walled complex of tents and makeshift structures that would become the British base camp and observatory. Cook had two weeks before he selected the sandy peninsula, dubbing it Point Venus. The promontory proved a strategic choice, being bounded on two sides by water and within cannon range of the *Endeavour*.[14] (Trying to keep up peaceful relations with the Tahitians, Cook nevertheless planned for worst-case scenarios too.) The explorers put up with some thievery of their metal implements and garments. But the big wooden box that evaporated into nothingness on May 1 was a different story. Without the astronomical quadrant the box contained, *Endeavour's* mission to Tahiti would be dangerously compromised. Fort Venus's superintendents hoped that the quadrant's thief wouldn't open the secure box and find the delicate triangular instrument within. Or at least they hoped the crook would discard the instrument, seeking out something more relevant to the island's daily life—or at least something sharper.[15]

On a sunny and breezy Tuesday morning, May 2, Banks took the situation into hand.[16] The crew and officers had already searched the whole of Fort Venus for the missing quadrant and turned up nothing. Banks started looking in the woods surrounding the fort. At a nearby river, the gentleman traveler ran into one of the local chiefs, Tubourai. "[He] immediately made with 3 straws in his hand the figure of a triangle," Banks recorded. "The Indians had opened the cases. No time was now to be lost!"

So Banks, Green, an unnamed midshipman, and Tubourai raced into the Tahitian jungle to pursue the thief who could singlehandedly scuttle Britain's greatest scientific expedition. The 91-degree heat was stifling, but they ran whenever they could. "Sometimes we walk'd, sometimes we ran when we imagin'd (which we sometimes did) that the chase was just before us," Banks wrote. Three miles into their seven-mile inland trek, Banks sent the midshipman back to Fort Venus to summon reinforcements. And so the sweaty, panting voyagers ran through black flies and mud, forest and clearing, till they reached the village where Tubourai had heard the quadrant had been taken.

The day was wearing on, and no doubt all three overheated runners were hungry and thirsty. But once they'd arrived, Banks level-headedly sat down among the reported "hundreds" of villagers who soon surrounded him. He'd learned some local customs that he now wielded to his own favor. Banks drew a ring in the grass and sat in the middle of it, ready to hold court. ("Mr. Banks . . . is always very alert upon all occasions wherein the Natives are concerned," Cook later observed.)[17] Grabbing the villagers' attention with his exposed but unbrandished pistols, Banks used his storytelling and explanatory skills to convey his words to Tubourai, his interpreter.

Piece by piece, the quadrant and its components began to emerge from the huts. Once the main body of the quadrant had been returned, Green looked it over to see what could be salvaged after its rough transit inland. Green found small parts of the instrument missing, some of which were returned—some not. But ultimately Green satisfied himself that what they had would suffice.[18] "We pack'd all up in grass as well as we could and proceeded homewards," Banks recorded. "After walking about 2 miles, we met Captn Cook with a party of marines coming after us. All were, you may imagine, not a little pleas'd at the event of our excursion."[19]

TAHITI

June 2–3, 1769

To scout for backup sites in case of cloudy weather, Cook had sent men to two other nearby locations—one to the east and the other to Moorea, a nearby westerly island. But they wouldn't be needed. The night of June 2 brought a tropical sunset unmarred by clouds—and calm to Fort Venus. "This day prov'd as favourable to our purpose as we could wish," Cook wrote the following day. "Not a cloud was to be seen the whole day, and the air was perfectly clear, so that we had every advantage we could desire in observing the whole of the passage of the planet Venus over the Sun's disk."

Cook posted sentinels around the observatory at Fort Venus, to ensure no thieving or other meddling by locals got in the way of job number one on that sunny Saturday morning. Earlier Green had set up a pendulum clock inside the officer's tent, erected right next to a wooden-walled, canvas-roofed portable observatory. Per his instructions, Green had fastened the clock to a firm wooden stand and set the pendulum to the familiar length already established for Greenwich.

The clock had been used before. It was on Nevil Maskelyne's unsuccessful 1761 Venus transit voyage to St. Helena and then again, in 1766–1767, as one of the main timekeepers for Charles Mason and Jeremiah Dixon's surveying trip to measure the border between Pennsylvania and Maryland—a line that immortalized the astronomers' names.

Outside the observatory building, Green and Cook each set up two-foot telescopes—as well as the repaired astronomical quadrant, with a wooden barrel serving as its makeshift stand.[20] The Royal Society mathematician and optical instrument maker James Short had made these Gregorian telescopes—named after a Scots mathematician from the previous century who'd proposed the scope's compact two-mirror design that had superseded Isaac Newton's pioneering reflectors. The pair of two-foot brass instruments stood on sturdy brass stands that nevertheless belied their delicate optics. It was a stroke of good fortune that no idle Tahitian hands had absconded with these finely crafted devices. A broken quadrant, in the hands of a good watchmaker like Spöring, could be refitted. A broken telescope might have been beyond repair.

Through the Gregorian reflectors' 4 3/4–inch eyepieces, dimmed at the front end by smoked glass solar filters, Cook and Green trained their eyes on the star that makes day.[21] At 7:21 AM and 20 seconds, Green was the first to see what planet astronomers all over the earth would be carefully viewing for the next six or so hours. Five seconds later, Cook shouted out that he spotted Venus too. Another twenty-one seconds after that, Solander recorded his first sighting of Venus's ingress.[22]

Here Cook, the preeminent tactician and military mind, was out-flanked by nature. He might have bested the odds by crossing an entire unwelcoming planet to reach his remote destination, but the same optical trick that had marred 1761 data fooled him too. "We differ'd from one another in observing the times of the contacts [of Venus with the edge of the sun] much more than could be expected," Cook wrote in his journal.[23] As the chief astronomer of the voyage, Green should have known what was coming. But Cook and Green collected their data as if the black drop effect was something unexpected and unknown.

During the transit, Cook and Green used special adjustable eyepieces (so-called Dollond object glass micrometers) that enabled measurements of Venus's apparent diameter. Both found the planet just under one arc minute in size—54.77 and 54.97 arc seconds, respectively. The same micrometer enabled Cook to measure a series of the angular distances between Venus and the sun's edge as the planet inched its way across the blazing solar face.[24]

And blaze the sun did on June 3. Cook noted in his journal that the mercury topped 119 degrees Fahrenheit by midday. "We have not before met with 119," Cook recorded with characteristic terseness. Observers the world over felt the sweat of pressure as Venus inched toward departing the solar disk for the second and final time of the eighteenth century. But at Fort Venus, nervous perspiration could hardly compete with the body's natural reaction to such oppressive heat.

At 1:09 PM and 46 or 56 seconds (Cook and Green, respectively) and then again at 1:27 PM and 45 or 57 seconds (Cook and Green, respectively), the mission's two sweat-soaked astronomers recorded the two final points of contact as Venus exited the solar disk. They later drew pictures of the plastic membrane that briefly appears to connect the two astronomical bodies at the moment of their edges making contact.

Green recorded all the expedition's numbers but postponed conducting any big calculations or other data reduction projects until a later time.

Meanwhile, on Moorea, an island nine miles northwest of Tahiti, Banks had joined Lieutenant Gore, Spöring, surgeon Monkhouse, and the surgeon's brother, midshipman Monkhouse, for a backup observation voyage in case clouds ruined Fort Venus's day. Records of the observations, instruments, and even observers conflict with one another.[25] Fortunately their data turned out to be unnecessary.

Banks, however, had spent the day chasing another Venus.

"At sunset I came off having purchas'd another hog from the king," the island's visiting gentleman recorded in his journal. "Soon after my arrival at the tent 3 handsome girls came off in a canoe to see us. They had been at the tent in the morning with [King] Tarroa; they chatted with us very freely and with very little persuasion agreed to send away their carriage and sleep in [the] tent—a proof of confidence which I have not before met with upon so short an acquaintance."[26]

Chapter 11

BEHIND
THE SKY

San José del Cabo
May 1769

The cool sunrise over the Gulf of California welcomed the French and Spanish voyagers to their new tropical home. The welcoming party, however, was slightly less impressive.

"We hurriedly began the same day the task of transporting our instruments and baggage to a small Indian town one mile from the beach," the Spanish observers Doz and Medina wrote, "which took three days because of the rough surf on the beach and the fact that there were no more than six Indians available to carry our effects to the mission—all the rest being ill from an epidemic prevalent in that town since the beginning of November."[1]

Their ultimate destination was an inland Franciscan mission, Misión Estero de las Palmas de San José del Cabo Añuití.[2] These missions operated under a simple system of exchange with the natives: we convert you to Christianity and then help to feed and clothe you—or God help you. God generally wasn't too helpful in these situations, especially with local immunities unaccustomed to old-world contagions—and not a few massacres of the native masses darkened the history of the period.

Once they'd arrived at Misión Estero, a new mood of urgency seized the travelers. Chappe wrote, "I made haste to establish myself at San José and to prepare for my preliminary observations. Myself and all my train took up our abode in a very large barn. I had half the roof taken off towards the south and put up an awning that could be spread out or contracted at will."

Chappe enjoyed ready access to the mission's facilities—even to tear into buildings—because there was no one to stop him. The present location of the mission was just sixteen years old.[3] Moreover, San José's Franciscan owners had been running the establishment for only a year and a half. The mission's Jesuit founders had, like all other Jesuits in New Spain, been expelled the previous February with extreme prejudice. The king of Spain left severe orders that if just one Jesuit—even if ill or infirm—was found in his New World domains, the viceroy of New Spain would be put to death.

To fellow Spaniards in southern Baja the orders from Madrid must have seemed bizarre. The Jesuit fathers on the peninsula were, by all accounts, upstanding practitioners of their faith, seeing their role on this earth as rescuing as many souls as mortal time permitted.[4] Nevertheless, the indigenous Pericu nation of southern Baja was already dubious of the missionaries. Polygamists who ceded religious authority to a hierarchy of *guamas*—witch doctors—the Pericu, as a rule, gave no quarter to Christ and his legion of saints. Instead, most Pericu lived in the hills, dropping by Misión Estero occasionally to sample its dates and grapes and begrudgingly to take, as one resident put it, "intensive refresher courses in Christianity." No wonder, then, that San José del Cabo boasted the smallest numbers of converts among the peninsula's eighteen missions. Still, ample ears remained to hear the sermons. For every human living at San José del Cabo, the stables counted as residents some three mules or horses—or, as the Pericu called the animals, "large deer."[5]

With syphilis running rampant among the Pericu for two generations, a second plague was hardly welcome. And yet, beginning in July the year before, a "contagious fever" had descended on Misíon Estero, spiriting away the mission's founding father. This *grande enfermedad,* which had shuttered the Santiago mission to the northeast, was an epidemic strain of typhus—"jail fever."[6]

According to one account of the period, jail fever "begins with a sensation of coldness and shivering, somewhat resembling the fit of an ague. . . . Soon afterwards, the patient complains of a pain in his head and back and sometimes in other parts of his body. [He suffers] of nausea and sickness at stomach, but he seldom vomits; of great lassitude, weakness and weariness; of dejection of spirits; of heat frequently alternating with the cold shivering fits; of thirst; and of loss of appetite. His sleep is also confused and disturbed with frightful dreams."[7]

And this is only the first stage of jail fever.

Misión Estero, San José del Cabo
May 19–28, 1769

The network of roads and mail routes between the eighteen missions on the peninsula was a feat of geography—if not scheduling. Even during a deadly epidemic, news and letters still ferried from hub to hub, infected or no. In Santa Anna, a mining town fifty miles northeast of Misión Estero, a royal officer of the Spanish crown, Joaquín Velázquez de Leon—an amateur mathematician and astronomer— had learned of the arrival of Chappe and his cohorts at San José del Cabo. Velázquez posted a letter asking to join the expedition and help conduct observations. Doz and Medina wrote back that Velázquez could indeed provide assurance and assistance. But such help consisted of Velázquez staying put. "Don Salvador de Medina and Don Vicente [de] Doz," Velázquez recalled, "replied that although they would take great pleasure in our concurring for the day of the observation, it would be better if I made mine in Santa Anna in case theirs failed because of cloudiness."[8]

Medina and Doz were thinking about other contingencies too. The jail fever spreading through their local workforce was as serious as death. "The numbers that were daily carried off too plainly showed the danger [we were] in," Chappe's assistant Pauly later wrote. Pauly recalled that once the expedition had made landfall and had begun to set up, Chappe became as focused as he'd ever been in his life. Although one-third of the local population had already died of the fever, he refused to consider relocating. "We might have escaped the contagion by going on to Cabo San Lucas [18 miles to the southwest], and this was what the Spanish officers proposed. But they were within a few days of the transit, and a second removal would have lost them very precious moments. Mr. Chappe, less apprehensive of endangering his life than of missing the observation—or making an imperfect one—declared he would not stir from San José, let the consequences be what they would."[9]

Another external force exerted its influence on the expedition, this time for the overall good. In the six months preceding their arrival, the inspector general of New Spain, José de Gálvez, had been ordering repairs to colonial properties in southern Baja and pressuring the Franciscans to improve the native population's lot. Gálvez knew that the French and Spanish visitors would be reporting back not only their astronomical data but also the local living conditions. Perhaps Chappe's emerging reputation as a noteworthy travel writer—his *Voyage en Sibérie* had recently been published in France—inspired Gálvez's cosmetic commandments all the more.

In late 1768, Gálvez wrote to a Franciscan official that he didn't want "a few learned strangers . . . [to] find in this province and its missions the wretched objects and horrid deserts which I encountered four months ago." The inspector general feared that these astronomers would return to the Old World and "publish in their narratives that the greatest and most pious monarch of the world is, in California, the lord of deserts and that he has as subjects Indians who go about as vagrants and live like untamed brutes." So Gálvez ordered extra food and clothing from the mainland and a contingent of soldiers to distribute the provisions and assist in ensuring Baja "look[ed] more prosperous and the people more civilized when the scientific expedition . . . should arrive."[10]

Those same (increasingly sickly) Indians constituted the local workforce. But Gálvez also imported workers from missions to the north to assist Chappe, Doz, and Medina in their fortnight-long buildup to June 3.

Doz and Medina opted for their own facilities nearby rather than relying on Chappe's site, which could also be a single point of failure. Both Spanish and French observatories had to perform the same essential functions that challenged their counterparts in Tahiti and Vardø.

The essential transit measurement was, fundamentally, one of duration. Chappe and his Spanish counterparts knew that Venus's disk would begin touching the solar face sometime around noon on June 3 and would complete its passage at approximately 6:00 PM. The timing of the actual event, however, needed to be accurate down to the second. For starters, this meant ensuring the pendulum inside the clock never felt even the slightest sway from Chappe or others coming and going from the observatory, accidentally bumping it, opening or closing doors, and so on.[11] Chappe noted that his observatory's floor was plenty firm, but now the clock needed an unmovable vertical brace. This meant a big piece of lumber.

The southern Baja peninsula was verdant with luscious fruits like pomegranate and prickly pear and abundant staples like rice and millet. But it had no great forests. Even the local brazilwood trees, which elsewhere in New Spain grew stout and tall, scarcely reached beyond human dimensions in the sandy and salty earth.[12] Fortunately, Chappe had already anticipated this need as well as the peninsula's deficiency in fulfilling it. In San Blas, before crossing the Gulf of California, Chappe had secured a plank of dry cedar for just this purpose. With the help of his assistants and whatever local labor could be mustered, Chappe sank and cemented his cedar plank two and a half feet into the ground, bracing it and the clock against one of his makeshift observatory's inside walls. He further buttressed the plank's two remaining sides and constructed a brick stand on the far side of the wall that braced the board and clock. Finally Chappe glued paper coverings atop a box that encased the pendulum clock's machinery, ensuring that neither wind nor dust could interfere with the clockwork apparatus.

While Chappe's clock was steadied to his converted barn's walls, Doz and Medina enjoyed no such luxury. They had to build their own makeshift observatory out of wood they'd brought with them from San Blas, plus local timber, cane, laths, clay, straw, and stones. The building measured an impressive forty-eight feet by sixteen feet, and had three holes in its roof. The holes, aligned along the path in the sky that the sun moves in early June, were each covered by a cloth. The flaps, Doz recorded, could be "raised and lowered with ropes, leaving open only space necessary for the telescopes in order to avoid the least movement the wind might cause."[13]

Windy southern Baja didn't affect just the design of the telescope housing. Securing their clock to unsteady walls would have subjected its delicate pendulum to extra jitters. And as Doz reported to his superiors, they found even the ground to be susceptible to vibration. So the Spaniards dug a hole in which they placed a pillar of rocks on which they then rested a five-inch-thick plank. The clock's cabinet rested on the plank and was vertically secured to another pillar of rocks behind it that was also sunk into the ground.

All clockwork seemed to be as independently suspended as circumstances would enable, and yet by May 28, Doz discovered his clock was drifting. The time between one solar noon and the next, according to the clock, was not twenty-four hours (as it should be on average) but closer to twenty-three hours and fifty-three minutes. It was ticking a little too slow; the pendulum rod was too long. Although the screw that allowed for fine-tuning the pendulum's length was worn down, Doz and Medina figured out a different way to shorten it. However, another couple days' observations revealed they'd shortened it too much. The timepiece soon recorded twenty-four-hour periods as being twenty-four hours plus another four to six minutes.[14] Their clock still needed fiddling.

As for the quadrants, telescopes, and other specialty equipment for observing stars and planets at their zenith, "piers of masonry were constructed to support the fundamental instruments," said a report about the two observatories. By May 31, both Chappe and Pauly's converted corn barn and Doz and Medina's hand-built observatory were as ready as they could be to observe a celestial phenomenon whose better acquaintance had led them into the heart of an epidemic-plagued territory on the far side of the New World.

Servants who had been perfectly healthy when Chappe, Pauly, Doz, and Medina had landed at San José del Cabo were now falling prey to the infection that still was creeping steadily through the population.

The second phase of jail fever, the contemporary literature said, leaves a sufferer "confined entirely to his bed on the second, third or sometimes on the beginning of the fourth day from the first accession of the cold shivering fit. Almost all the symptoms now increase. The pain of the head becomes acute. The debility is such that the patient cannot be raised to an erect or even sitting posture for any length of time without great danger of bringing on a fainting fit; the appetite is totally gone; the thirst increases."[15]

Misión Estero, San José del Cabo
June 3, 1769

Eight years before, Chappe slept on a mountaintop observatory in remote Siberia and, on the morning of June 6, 1761, recorded some of the most accurate Venus transit data of any observer in the world. Now, as the pale pearl dawn on June 3, 1769, roused the voyagers, a new day of cosmic inspection awaited. This time, no clouds threatened to steal the moment away.

During his seventy-seven-day passage across the Atlantic Ocean six months before, Chappe had already begun rehearsing his regimented procedures for transit day.

"I was busy," Chappe recorded in his journal, "during the crossing from Cadiz to Vera Cruz, calculating all the details of the passage of Venus for San José, putting together all the observations I was to make, arranging them so that none detracted from another, posting beforehand the site and disposition of each instrument, according to the operation for which I intended it. I drew up as well a master chart where all details of the observation were exposed in their order, and I fixed it the day before on the wall, facing my instruments, so that I might at every moment be able to recall what I was to do or to prepare for."[16]

Chappe's converted barn and Medina and Doz's clay hut, alive with anticipation and activity on this cloudless morning, sounded a bright grace note in a time of dirges. Even as families wept over dead spouses and children, and tribal healers gave whatever comfort they could to a gravely ill nation, two unyielding outposts pressed on with their mission. Even typhus could not close the window the skies were about to open.

In Chappe's barn, a servant counted out each second, while Pauly kept track of the passing of each minute and recorded Chappe's time stamps and observations. Through his eyepiece and through the smoked glass at the objective lens on the three-foot telescope, Chappe watched, as he put it, "Venus making a small indentation on the edge of the sun, perfectly defined." The time was 11:59 and 17.03 seconds.[17]

Chappe had initially considered enlisting Pauly to perform a redundant set of observations from another telescope inside the barn. But eventually Chappe decided it was better to concentrate his group's work on a single measurement. There were already redundant efforts on the mission grounds and at Santa Anna.

Doz and Medina had made the opposite decision. Each of the two chief Spanish observers would be looking through his own telescope and tracking Venus's motions independently. "The sun was nearing its zenith, and the telescope was therefore almost perpendicular," Doz recorded in his journal. "The position I found least inconvenient in observing the first contacts was lying flat on the ground."[18] Doz had calculated Venus would first be appearing on the sun's northeast edge, some 25 degrees above the solar equator. He recorded difficulty making out the exact moment of contact, however. By Doz's figuring, Venus first touched the sun's edge at 11:59 and 14 seconds. Medina independently observed 11:59 and 18 seconds. The second contact, that instant when the entirety of Venus's shadow first crosses over into the solar disk, came another 18 minutes and 11 seconds later—for Medina it was 18 minutes and 12 seconds, for Chappe, 18 minutes and 9.84 seconds.

Both Doz and Chappe recorded something else that the Russian polymath Mikhail Lomonosov—whom Chappe had met in St. Petersburg—had discovered in 1761. "I noted what I had hoped for days to discover," Doz wrote, "to find ... around the planet a faintly illuminated penumbra, which I took for its atmosphere."[19] As Venus cut a small semicircular hole into the edge of the sun, in other words, Venus's atmosphere scattered some of the sun's light and created the illusion of a faint halo surrounding the otherwise invisible remainder of the planet.

During the next six hours, Chappe labored furiously to measure every step of Venus's linear passage across the sun's face. Six times he used the telescope finder's crosshairs to measure the planet's angular size down to fractions of an arc second. Another twenty-eight times he measured the slowly changing distances between various edges of Venus and various edges of the sun. At 12:25 and 13.7 seconds, for instance, Chappe found the northern cusp of the planet was 2 arc minutes and 48.04 arc seconds distant from the sun's northern edge.

Finally, at 5:54 PM and 50.31 seconds, a sweat-soaked Chappe called out to Pauly that Venus was beginning to exit the sun's disk. Doz and Medina both recorded this initial "egress" at 5:54 and 47.5 seconds. Each said good-bye to the "morning star" as it took 18 minutes and 28.18

seconds (Chappe) or 17 minutes and 54 seconds (Doz) or 17 minutes and 59 seconds (Medina) to exit entirely.

Doz reported difficulties in seeing the exact time of Venus's exit due to what would later be called the black drop effect—the same optical illusion that was well reported on from 1761. The molasses-like drawing out of the moment of final contact was certainly no surprise, and yet Doz's notes suggest he hadn't prepared. The "shortcoming" in his data, Doz grumbles in his journal, "was caused by lack of time to establish the observatory on high ground and the decision prompted by the zeal of Mr. Chappe not to continue to our original destination, Cabo San Lucas."[20]

Doz's two explanations conflict with each other. One says there was not enough time; the other says there was plenty. The unmentioned factor, however, was the typhus striking down able bodies all around them. Doz's irritation may have been more deeply seated than he realized. For even as the observers were performing final calibrations and measurements in the transit's immediate aftermath, it struck.

Pauly wrote, "On the 5th of June, two days after they had observed the transit of Venus, Mr. Doz, Mr. Medina and all the Spaniards belonging to them, to the number of eleven, sickened at once. This occasioned a general consternation; the groans of dying men, the terror of those who were seized with the distemper, and expected the common fate, all conspired to make the village of San José a scene of horror."[21]

Chapter 12

SUBJECTS AND DISCOVERIES

Trondheim, Norway

September 2–13, 1769

The altar bread had grown moldy since Sajnovics and Hell were last in town. And with Communion supplies imported from the Netherlands, the priests in this Norwegian port city had no holy host to spare. The Jesuit astronomers had spent all of July and August surviving various hair-raising sea adventures on the 850-mile journey from Vardø. They relished a rare moment when their greatest worry concerned spoiled wafers.

On the morning of Saturday, September 2, Sajnovics sat in his unheated guest quarters and scratched out a letter recounting the sea journey to his Hungarian Father Superior. The soldiers and townsfolk in Vardø, Sajnovics said, "were not too thrilled to see us go. . . . They were forced to stay in this place nobody liked and everybody wanted to leave, wishing to be freed so much that they utter hundreds of sighs."[1] But the captain said the wind was finally favorable, and the astronomers had done all the astronomy they needed to do. So with a new pet ("the little fox we had purchased from a Finn in Vardø kept scaring people with its cute barking") and a new ship ("its cabin was sumptuously furnished, but it was small, and it had no stove"), Hell and Sajnovics and their crew were soon on their way.

"Only he who has experienced the unspeakable wrath of the Arctic Ocean and its many rocks and cliffs reputed for sinking countless ships can have an idea about the many dangers we had to face," Sajnovics wrote. Within a week, they'd already begun to hear about competing arctic transit expeditions. The Danish king had also sent, as a backup mission, astronomer Peder Horrebow to a Norwegian coastal town two hundred miles north of Trondheim. But Horrebow, Sajnovics reported, "was unable to see the transit because of the unfavorable weather conditions." The following week, Sajnovics had come within hailing distance of an English ship bearing two sets of astronomers who had tried to observe the Venus transit from another northern Norwegian location. "As we learned later on, neither of them saw Venus because of that impenetrable fog that was blocking the view on the 3rd and the 4th of June," Sajnovics wrote.[2]

"Oh amazing divine providence," Sajnovics wrote, "that from all these people who had prepared and worked hard, [God] only granted Father Hell the privilege to reach the goal for which so many have strived and hoped for."

During their Trondheim stay, Hell remained busy. He accepted an honorary diploma, awarding him in absentia membership in the Danish Academy of Sciences; the consul in Trondheim hosted the visiting dignitaries for a concert in their honor; the scientifically minded bishop shared his natural history collection with Sajnovics and offered to give the holy father a microscope for better conducting his research.

Then, on a clear and crisp mid-September morning, Hell and his team set off over the terrible roads out of Trondheim for the inland part of their journey. Like the sea voyage they'd just completed, the trip to Oslo would be much like the previous year's odyssey—only in reverse.

However, this time some of the locals now knew about the Venus transit expedition and were eager, as the scientists returned south, to inquire how it went. The night after Hell and Sajnovics set out from

Trondheim, they took dinner in a tiny hamlet where the locals excitedly asked about the astronomical mission, as they understood it at least. They'd learned from Horrebow and his assistant, when the Danish observers were passing through, that astronomers had been "searching for a lost star." So, Sajnovics recorded, the locals wondered if Hell and Sajnovics's destination might have been where the missing celestial object was hiding.

"The peasants ... were asking very seriously if we found the star that had been lost and that we had been searching for," Sajnovics wrote that night. "When we answered with 'Yes!' they were very grateful and told us that the other professors did not manage to find it."

Copenhagen
October 1769–May 1770

Denmark's King Christian VII had curbed his manias somewhat on his 1768–1769 grand tour of England and the Continent. He had indeed returned to Denmark a healthier and ostensibly saner man. Courtiers who had promoted the king's travels climbed the ladder of preferment, while skeptics who had doubted the king's capacity for self-healing were shunted aside. Christian VII's court—where a soupçon of intrigue was served up with every meal—had returned to Denmark in all its familiar forms. But outside the Amalienborg royal palace, the outlandish cost of the royal tour—the expensive gifts His Majesty handed out at every stop, the elaborate costume balls employing as many as 1,500 carriages—had tainted any feelings of joy that even Christian's supporters might have felt upon his return.

Such was the poisonous political atmosphere that Hell and Sajnovics entered as their train of carriages and baggage carts caught its first glimpse of Copenhagen on Tuesday, October 17. At their first official reception with members of King Christian's court, two days later, Sajnovics noted that minister of state, Count Otto Thott, "was glowing with joy and happiness for a whole hour. He was expecting that Father

Hell would realize a complete renewal of Danish astronomy." If the economic upheaval in the Danish and Norwegian population had risen to the travelers' attention, none chronicled it. And Sajnovics remained dazzled by the gilded glow of munificence. "It is unbelievable how much money this royal court directs toward the development of the sciences," Sajnovics wrote.[3]

Hell and Sajnovics carried with them data that other countries—and other Danes—had tried in vain to get for themselves. The transit expeditions to sites near the equator made the Hungarians' results worth all the more. Without good arctic transit observations for comparison, no amount of good luck and great brilliance by British and French equatorial explorers could make up for the fact that they were only performing one-half of the full, worldwide experiment. Hell was beginning to realize that fortune had handed him the key to the other half. In letters written from Copenhagen, Hell said that his employers prohibited his disseminating any information about the observations before he'd formally presented a report to Christian VII.[4]

Jérôme de Lalande in Paris began to process the data coming in, without access to Hell's numbers. Hell and Sajnovics holed up in Copenhagen for the winter, working on the complete account of their travels that they would then publish for the king. A spirit of international cooperation had graced nearly all transit missions to date—even in 1761 when sponsoring countries were at war with one another. But Hell and Sajnovics constitute the most glaring example of the proprietary extreme to which researchers and their sponsors sometimes take their prized results.

As Sajnovics wrote, for instance, Danish officials had "found out that our journey had been made public by the newspapers in Vienna. The secretary [of the Austrian embassy] said he wrote to [Austrian prince] Kaunitz and told him not to let this happen in the future."[5] By contrast, both Sweden and Russia promptly communicated their (inferior) arctic transit observations to the global clearinghouse for all Venus transit results, the desk of Jérôme de Lalande in Paris. Using the less than ideal arctic observations, French astronomers ventured first estimates of the earth's distance to the sun. Unsurprisingly, their initial efforts were off the mark. In January, for instance, Lalande calculated 90,500,000 miles—97.7 percent of the actual distance. They could do far better. Visionaries like Edmund Halley had in 1716, for instance, argued that the Venus transit could enable science to trace out a map of the solar system accurate to 99.8 percent or better.

Meanwhile, as the months passed by, Sajnovics wrote about taking lunch with prominent Danes like Count Thott, Count Bernstorff, and the court astronomer Christian Horrebow—brother of the transit voyager Peder. And waiting. "Hell finished his lecture series about the observation of Venus," Sajnovics wrote in December. "However, the editorial works related to the [book] got stuck."[6]

The incessant delays may have originated in the tectonic shifts taking place under every courtier's feet. As a three-day Christmas blizzard whited out the capital, Copenhagen descended into political darkness. King Christian's continental "cure" was proving to have been anything but. His Majesty's fickle moods and foul tantrums had begun to return, and as a result the balance of power at court was blowing capriciously back and forth like a wind vane in a tempest. "When [the king] is dressing, he may sit whole hours and more quite quiet, with eyes fixed, mouth open, head sunk, as though insensible," one courtier recorded at the time. "I knew him, and I have not forgotten that attitude, which always foreboded some violent scene and some revolution which is then being thought out."[7]

Sajnovics's prime sponsor at court, Count Bernstorff, had begun to find that despite all his influence and savvy, the king's private doctor, Johann Friedrich Struensee, was beginning to eclipse everyone.

The lowly physician would, astonishingly, reach an apotheosis surpassing all but the king himself—ending the careers of lifelong advisers like Bernstorff. Struensee would meet his own tragic end in 1772, when his love affair with Queen Caroline Matilda would prove to be one step too far in the doctor's personal quest for power.

Blissfully ignorant of the operatic level of conspiracy and intrigue going on all around them, on February 8, Hell and Sajnovics presented their long-awaited book, *Observatio Transitus Veneris Ante Discum Solis Die 3 Junii Anno 1769*, to the king.

Christian VII had received his personal copy of the report "with an indescribable gracefulness," Sajnovics wrote in a letter from Copenhagen. "This meeting resembled a scientific discussion that lasted for half an hour and the King was inspired to declare that he had understood Hell's contribution and its value, and he fully appreciated it. . . . He was immensely satisfied."

"We talked for about half an hour about the northern lights, the flood tide on the sea, and the peculiarities of the Lapp and Hungarian languages," Sajnovics wrote the night of his meeting with the king. "We even touched on the issue of the squaring of the circle. It seemed he had been well-informed about Hell's activity and he assured us that we fully reached his expectations and those of the others."

At last, Hell and Sajnovics's data could be shared with the scientific world. But, because of the eight- or nine-month delay, some scientists had their doubts about the data's veracity. A member of the Académie Française, Cardinal Paul d'Albert de Luynes, wrote to Hell in June about the whispers he'd been hearing that some thought the Vardø team's delay "could give rise to suspicion, that having had the time to receive other [Venus transit] observations, you could accommodate them with yours."

Others stood up for Hell and Sajnovics, though. The Swedish astronomer Anders Planmann confessed in a letter to a colleague that he was initially dubious of the dataset in question. But, Planmann added, "I free him from all suspicions concerning the correctness of his observation. . . . The circumstances which emerge from the report could not possibly have been fabricated."[8]

Others still had grown not suspicious but just angry.

On April 3, Sajnovics wrote, "We received a letter from Lalande in which he inquired, in a very arrogant manner, why we were so late in sharing the results of our observations with the people in Paris, adding some abominable threats to his questions. Finally he was asking about the Danish Scientific Society—as if it was something unknown."

Three days later at the Danish Scientific Society, Horrebow presented an account of his unsuccessful Venus transit voyage to Dønnes, Norway. "They did not read the letter in public that Lalande had addressed to the society," Sajnovics noted.

The voyagers, having worked so hard to procure royal favor, spent a month enjoying Copenhagen with doors swung wide open for them. At the Cathedral, for Palm Sunday and Easter mass, they were seated among dignitaries. The Danish Astronomer Royal gave them a tour of the king's library, perusing modern and ancient biblical texts and the manuscripts of court astronomer Tycho Brahe. They even spent a day, April 19, onboard a Danish warship to take a full tour and sumptuous lunch among captains of the fleet. "Music was constantly played during lunch," Sajnovics wrote. "Meanwhile the commander-in-chief was talking about how he had suffered while he was kept captive by the Moroccan Emperor, and how the latter stabbed him in the chest with a sword, and how the emperor shot at the bottle in his hand."

On May 10, King Christian gave Hell a farewell gift—a portrait of King Christian. Five days later, as the voyagers were beginning to pack for their final return to Vienna, Sajnovics handed over to a local official his own magnum opus, a book announcing the discovery of the surprising similarities between the Hungarian and Lapp languages. Sajnovics's *Demonstratio Idioma Ungarorum et Lapponum Idem Esse* is still celebrated today as an early triumph in comparative linguistics.

"We said goodbye to a few acquaintances," Sajnovics wrote on May 19. "We had lunch at Count Thott's place in the company of four Excellencies. After lunch they showed us two triumphal arches—under each there was a telescope positioned toward the Sun as well as two beautiful cottages—representing the victorious return of astronomy."

As far as Sajnovics was concerned, the two Hungarians accomplished their goals completely. Not only had they successfully observed the Venus transit, but Sajnovics felt other scientific observations the pair performed during the trip—concerning the earth's magnetic fields and the northern lights—might be equally important too. Sajnovics wrote in a letter from Copenhagen that Hell's book, *Observatio*, "presents 12 chapters . . . the mere titles [of which] already raised the expectations of the scientific circles from here, which is proof that we spent our time in Vardø working very hard and not with relaxation, and we used that time with result for the sake of science."[9]

Chapter 13

SAIL TO THE SOUTHWARD

SOUTHERN PACIFIC OCEAN AND EASTERN
COAST OF NEW ZEALAND

August 10–November 9, 1769

The Royal Society's original choice to captain *Endeavour*, Alexander Dalrymple, never set foot aboard the refitted ship that would circumnavigate the planet and help chart the entire solar system. But Dalrymple still had pull. His legacy endured into the *Endeavour*'s after-mission. Upon completing the Venus transit observations and requisite measurements of moon and stars to get a precise fix on Fort Venus's latitude and longitude, Cook opened a sealed set of orders from the Admiralty.

"So soon as the observation of the planet Venus shall be finished," the orders stated, "you are to proceed to the southward in order to make discovery of the Continent above mentioned [Terra Australis Incognita] until you arrive in the latitude of 40° [south], unless you sooner fall in with it."[1]

Discovering the fabled southern continent was one of Dalrymple's lifelong passions. And the Scottish hydrographer worked to ensure that

Endeavour would do all in its power to reach its shores, whether he was onboard or not.

Cook never believed in the missing continent, but orders were orders. So *Endeavour* spent a final month at Tahiti—where on the eve of the ship's departure two marines had decided to stay with their new "wives." Cook had them brought back aboard ship and flogged. And then Cook rounded the archipelago and headed due south on August 10.

Banks brought a Tahitian onboard to help guide the ship through the South Pacific. Tupaia, an island chieftain, befriended the voyagers and while Cook would not personally be responsible for the foreign traveler, the captain recognized the value of an indigenous envoy who could accompany the ship. Banks wrote, in the characteristic paternalistic tones of his age, "I do not know why I may not keep [Tupaia] as a curiosity, as well as some of my neighbours do lions and tigers at a larger expense than he will ever put me to. The amusement I shall have in his future conversation and the benefit he will be to this ship, as well as what he may be if another should be sent into these seas."[2]

On discovering the island Ohetiroa (today Rurutu) on August 14, Cook dispatched Tupaia, Banks, and Lieutenant Gore in the pinnace to land and "speak with the natives and to try if they could learn from them what lands lay to the southward of us," Cook wrote.[3]

Having spent the day ashore, Banks appreciated all the more the Tahitian paradise he'd recently bid farewell to. "The island to all appearance that we saw was more barren than anything we have seen in these seas," Banks journaled. "The people seem'd strong, lusty and well made."[4]

As the pinnace approached land, islanders waded out to greet the visitors. But the scene quickly devolved into a bad bazaar experience, with grabbing at anything hands might reach. Marines fired their muskets into the air—not intending to harm. Nevertheless, one islander sustained a small head wound. The away team soon returned to *Endeavour*, and Cook decided to weigh anchor.

"It appear'd that we could have no friendly intercourse with them until they had felt the smart of our fire arms, a thing that would have been very unjustifiable in me at this time," Cook wrote. "We therefore hoisted in the boat and made sail to the southward."

By September 2, *Endeavour* had reached the parallel her orders demanded. "At 4 p.m., being in the Latitude of 40° 22 [arc minutes] south and having not the least visible signs of land," Cook wrote, "we wore and brought to under the fore sail and reef'd the main sail. I did intend to have stood to the southward if the winds had been moderate so long as they continued westerly . . . but as the weather was so very tempestuous, I laid aside this design, thinking it more advisable to stand to the northward."[5]

And so died the Southern Hemisphere's great "lost continent," as unceremoniously as a shift in the day's breeze.

Endeavour spent September and early October charting an inverted V—first bearing northwest, then southwest—toward what is today called New Zealand. On October 7, a teenage surgeon's assistant spotted land. Cook rewarded one Nicholas Young a gallon of rum.

Could Young Nick's Head[land], a place-name that remains to this day, be the first hint of the legendary Terra Australis? Cook and Banks were, as ever, dubious that the mythic continent existed. But Cook couldn't know for certain what kind of landmass he was dealing with until he could circumnavigate it. (*Endeavour* ultimately would. Young Nick had in fact lighted upon the eastern coast of New Zealand's North Island.)

The next ten days turned bloody. Shore parties encountering New Zealand tribesmen, Maori, violently ended their brief visit three times—each with fatalities—when Maori tried to steal officers' belongings or kidnap Tupaia's young assistant. Cook ordered *Endeavour*'s course reversed, and she began sailing northward along New Zealand's eastern shore instead.

On November 4, *Endeavour* pulled into a bay near the North Island's northern tip.

"We were accompanied in here by several canoes who stay'd about the ship until dark," Cook wrote. "And before they went away they were so generous as to tell us that they would come and attack us in the morning. But some of them paid us a visit in the night, thinking no doubt but what they should find all hands asleep. But as soon as they found their mistake they went off."

Cook hoped to find a sheltering harbor to observe the more frequent—much less astronomically useful—transit of Mercury on November 9. Astronomers in Europe would also be performing careful observations of the planet Mercury crossing the sun's disk. And while Mercury's transit is too brief and the planet's solar distance too close to assist in measuring the solar system's dimensions, its November 9, 1769, transit was still a universal time stamp. Cook's observation of the Mercury transit would at least ensure that he knew the exact longitude and latitude of "Mercury Bay," as Cook had now dubbed the inlet. All other longitudes and latitudes of Cook's (stunningly accurate) New Zealand map would build from the Mercury transit longitude.

As Banks recorded in his journal, he continued collecting specimens while Cook performed his astronomical and navigational tasks.

"At day break this morn a vast number of boats were on board almost loaded with mackerel of 2 sorts, one exactly the same as is caught in England," Banks wrote. "We concluded that they had caught a large shoal and sold us the overplus what they could not consume, as they set very little value upon them. It was however a fortunate circumstance for us, as by 8 o'clock the ship had more fish on board than all hands could eat in 2 or 3 days. And before night so many that every mess who could raise salt, corn'd as many as will last them this month or more.

"After an early breakfast the astronomer went on shore to observe the transit of Mercury," Banks continued. "Which he did without the smallest cloud intervening to obstruct him, a fortunate circumstance as except yesterday and today we have not had a clear day for some time."

OFFSHORE OF NEW HOLLAND (AUSTRALIA)
June 11–26, 1770

Captain Cook didn't know it at the time, but when he swung into his cot on the night of Monday, June 11, 1770, *Endeavour's* wonderfully unsexy flat keel was about to save his entire mission. At 11:00 PM, even as recent soundings of the ocean depth had read 17 fathoms (31 meters), the ship suddenly and without warning crashed onto a coral reef. The cacophonous crunch of splintering hull certainly sounded dire at the moment of impact. But practically any other ship of *Endeavour's* size in His Majesty's Navy would likely have been doomed to sink. As it was, *Endeavour* had a blind date with the Great Barrier Reef and might still be salvaged.[6]

The ship had circumnavigated all of New Zealand, creating exquisitely detailed charts of the two islands that made New Zealand's for a brief time the most accurately mapped coastline in the world. Only after *Endeavour* had completed its mission and Cook's new and innovative cartographic methods were applied to European coastlines did Cook's New Zealand map have any true rivals.[7]

Now she was plying the east coast of New Holland (today's eastern shores of Australia) and discovering new worlds of flora, fauna, and native populations that kept its gentlemen explorers working day and night. One fishing expedition on *Endeavour's* yawl—the smallest of the ship's three launchable sailing vessels—had recently returned with six hundred pounds of stingrays. Over the ensuing fortnight, Banks's young artist Parkinson made ninety-four sketches of various flora samples collected from shore visits along the way.

The Australian continent's cornucopia of natural wonders was proving so exciting—and distracting—that some warning signals might have been ignored. No European expedition had come this close to the Great Barrier Reef before. When the French explorer Louis Antoine de Bougainville had sailed through the same waters two years before, he heard the roar of surf breaking on the nearby reef. "This was the voice of God," the French mariner recorded in his journals as his ship fled toward open ocean, "And we obeyed it."[8]

After the jolt of impact, Cook raced up to the main deck in his underwear. He ordered all sails taken in to avoid being blown farther onto the reef. And the surf that Bougainville had discovered now broke onto *Endeavour*'s hull, momentarily lifting and then dropping the boat with each passing wave. Thankfully, the moon was nearly full, and all hands on deck could see what they were doing as the captain assessed the dreadful situation. Within minutes, the crew watched helplessly as shards of *Endeavour* floated away from the ship and into the reef's shallow waters.

Cook ordered a detachment of men onto the longboat to attempt to pull *Endeavour* back off the reef complex. The bark would not budge. Engineers reported from below deck that the ship was taking in water and needed to be refloated fast lest the lowering tide strand her till the higher waters returned again the following evening.

Cook delegated his officers to quickly assess what ballast could best be tossed overboard to lighten the ship's load. The crew jettisoned casks, drinking water, spoiled food, iron and stone, and six cannon. *Endeavour*, now forty or fifty tons lighter, still wouldn't move.

Further salvaging techniques failed over the coming day, until finally at 9:00 PM, the ship righted itself. She might make it to shore, Cook hoped, if only her wounds could be staunched. The ship's master had found a harbor to the north. So *Endeavour* plodded ahead in its new race against the clock. To slow the leakage of seawater into the ship, midshipman Monkhouse filled a spare sail with wool, animal dung, and tar-filled rope fibers. They lowered the "fothering" overboard from the main deck, and like a drain plug in a bathtub, the water's suction power fastened the makeshift patch against the broken hull.

Endeavour made it to the mouth of a river—a waterway Cook named Endeavour—where at high tide the bark was beached so the blacksmith and carpenters could begin repairing the wounded vessel. The reef itself, an inspection soon revealed, had saved the entire ship from sinking. The fothering had only closed up part of the hull's hole. Part of the hull-cutting reef had broken off, and the fothering had inadvertently wedged the coral into place to fill the gap.

By June 22, all crew members were camped onshore and recuperating from the nearly fatal disaster. Banks, of course, seized the opportunity to collect more specimens. Along with the usual catch of birds and satchels full of plants, Banks and Cook caught glimpses of a beast that was nothing like they'd ever seen before.

"It was of a light mouse color and the full size of a greyhound," Cook recorded. "I should have taken it for a wild dog but for its walking or running in which it jump'd like a hare or a deer."[9] According to Tupaia their Tahitian interpreter, whose interpretive skills were weakening as the ship ventured farther and farther away from his native island, the locals called this animal "kangooroo."

Banks now turned his sights on hunting the strange animal. The newly discovered marsupial, Banks wrote, "hop[s] upon only its hinder legs, carrying its fore close bent to its breast. In this manner, however, it hops so fast that in the rocky bad ground where it is commonly found it easily beat my greyhound, who, though he was fairly started at several, killed only one and that quite a young one."[10]

Cook concluded, simply, that the kangaroo "proved most excellent meat."[11]

BATAVIA (JAKARTA, INDONESIA)
October 1770–January 1771

The repaired *Endeavour* continued to leak and ultimately required the attention of professional shipwrights in a bona fide naval yard. Cook knew his best chance of surviving passage through the Indian Ocean and beyond the Cape of Good Hope was to refit *Endeavour* at the Dutch East India Company's headquarters in Batavia. A ship that probably couldn't have made the return voyage thus docked at the company shipyards. In Banks's words, crew members were "rosy and plump" when the ailing ship pulled into port.

The Dutch had built Batavia in the early 1600s and, naturally, filled the city with canals. Over the ensuing century and a half, however, the stagnant water, sewage, and animal carcasses in the canal made the East India Company capital city a festering swamp of malaria and dysentery.

During the twelve weeks the shipwrights required to repair *Endeavour*, all but one of the ship's crew fell sick at least once. *Endeavour*'s healthiest shipmate turned out to be John Ravenhill, a sail maker characterized by Cook as a drunk. By Cook's estimate Ravenhill was "an old man about 70 or 80 years." Actually Ravenhill was forty-nine.[12]

On December 26, when Cook finally weighed anchor and set sail for Cape Town, the captain wrote that *Endeavour* left Batavia "in the condition of a hospital ship. [We lost] seven men, and yet all the Dutch captains I had an opportunity to converse with said that we had been very lucky and wondered that we had not lost half our people in that time."[13]

Cook would continue losing men on his "hospital ship" as they set sail across the Indian Ocean. (The unhealthy "fresh" drinking water that *Endeavour* took aboard at Batavia was the likely culprit.)[14] During the first two weeks out of Batavia, another seventeen died. Banks's young artist, Parkinson, died on January 27, 1771. Two days later, Cook recorded in his journal the final passage of his chief astronomer.

"In the night died Mr. Charles Green who was sent out by the Royal Society to observe the transit of Venus," Cook wrote on January 29. "He had long been in a bad state of health, which he took no care to repair—but on the contrary lived in such a manner as greatly promoted the disorders he had had long upon him. This brought on the flux which put a period to his life."[15]

Chapter 14

ECLIPSE

Misión Estero, San José del Cabo
Summer 1769

The mildly good news, as some of those suffering from the Baja fever began to see, is that ultimately the plague was survivable. "The patient talks somewhat incoherently, yet knows his friends and will answer questions with tolerable distinctness," the jail fever chronicler said about the third phase of the illness. "In this situation he continues six, seven, eight, nine, ten or eleven days and from which, if the symptoms do not increase, he gradually recovers."[1]

By June 5, Chappe had become both chief astronomer and chief doctor to the mission. "Mr. Chappe had brought with him from France a little chest of medicines and some physic [medical] books," Pauly wrote. "In this emergency he was an occasional physician. He examined the symptoms of the disease, then consulting his books, he endeavored to find out the proper remedies. But he soon found himself as much at a loss as those who formerly consulted the oracles, whose ambiguous answers frequently admitted of two opposite meanings, and left them as much in the dark as before."[2]

Chappe remained busy in his observatory, too. On the night of June 6, he trained his telescope on Jupiter to mark the time its moon Io crossed behind the planet.[3] Chappe notes in his logbook that his observation of Io's eclipse behind Jupiter was something close to ideal. "Perfect observation," he wrote.[4] The next four nights Chappe was up late observing the "culmination" of the stars Arcturus and Kornephoros (the second-brightest star in the constellation Hercules). An astronomical culmination is the equivalent of solar noon for any given star. It's the highest point in the sky that a star's east-west journey carries it on any given night. And just like regular noon, when a star is at its culmination, it's almost exactly halfway between where it rose in the east and where it'll be setting in the west. This makes two different, nearly perpendicular angles on the sky.[5] So it provides an opportunity to test slight errors in one's astronomical quadrants. On the night of June 7, Chappe measured both the distance toward the eastern horizon from Arcturus and Kornephoros at culmination. On June 8, he took the same measurement but this time toward the western horizon. Later analysis proved that the quadrant he used to take transit data carried, over wide angles, an intrinsic error of 1 arc minute and 25 arc seconds.[6]

During the 9:00 AM and 2:00 PM hours on June 8, Chappe busied himself with careful measurements of the sun to test any slight or subtle drift rates in his pendulum clock. And then on the nights of Friday, June 9, and Saturday, June 10, he went back to using his quadrant to take culmination measurements on Arcturus and Kornephoros again.

For Chappe, Sundays at the mission were hardly days of rest. Sunday, June 4, the day after the Venus transit, Chappe had been busy both day and night measuring the sun's motion and Jupiter's moons to continue collecting data that would pin down both his clock's exact error rate and his observatory's longitude. But the following Sunday, June 11, was a different story. This time, as Pauly noted, Chappe "had a violent pain in his side and was delirious at times."[7] Typhus had struck the group's leader.

Bouts of fever and delirium came in twenty-six to twenty-eight-hour stretches. "He was forced to prepare his own medicines," Pauly noted. When he was still well, Chappe had relied on another healthy member of the mission to mix the tonics. But one vial was mistaken for another, and as a result the group's feverish artist Noël had nearly been poisoned. Chappe wasn't going to make that mistake again.

Chappe's astronomical measurements, of course, stopped. But Chappe also knew that on June 18, one week into his illness, there would be a lunar eclipse. Astronomers all across Europe would be timing the exact moment of the eclipse's beginning and end. This was the ultimate cure-all for his longitude problem. Chappe had resolved, regardless of his health, to measure the June 18 eclipse with the same precision that had defined his transit data. "It is inconceivable how Mr. Chappe, low as he was, laboring under his malady, weakened by the fever fits he had gone through, could lend as close an attention to this phenomenon as the ablest observer could have done in full health," Pauly wrote. "Indeed he had much ado to hold out to the end of the observation. He was taken with a fainting fit, and a pain in his head. . . . He desired to be let blood; his interpreter, a surgeon who had never practiced much, and who was himself sick, tried to bleed him but missed. However, encouraged by Mr. Chappe, he tried again and succeeded."[8]

As Pauly observed, "It will be a matter of admiration to look over the account of this observation." The moon entered the earth's penumbra (half shadow), Chappe, recorded, at 10:45 PM. At 11:08, Chappe recorded, "the eclipse started, I think, within the minute. The shadow is so clear and the moon is so brilliant, I think I estimated that start time too late."[9] So Chappe picked up the pace. Starting at 11:11 PM and 41 seconds, he began to record the entry of the earth's complete shadow across every lunar crater he could find. He recorded a stunning thirty-two observations, revealing not only expert offhand knowledge of lunar geography but also a phenomenal stamina in the face of a severe malady.

As the eclipse passed its halfway point, and the end of the earth's shadow was sweeping its way past the same lunar craters, starting at 12:40 AM, Chappe recorded another thirty-four time-stamped observations as the "lighted segment" of the eclipse swept past the Harpalus, Aristarchus, Galileus, Menelaus, and Tycho craters. No doubt shivering and gripped by extraordinary pain, Chappe recorded his final entry for the night at 2:48 AM. "The edge of the moon is perfectly [out of eclipse]," Chappe wrote. "But it still seems smeared."[10]

Ultimately Chappe's lunar eclipse data was not deemed as useful as his other longitude-determining observations, including both observations of Jupiter's moons and observations during the Venus transit. But Chappe's data in aggregate yielded a longitude determination for Misión Estero of 112 degrees, 2 arc minutes, and 30 arc seconds west of the royal Paris observatory—or 109 degrees, 42 arc minutes, and 19 arc seconds west of the Greenwich prime meridian.[11] By any measure, this was an impressive result.

Misión Estero was shuttered the following century, and during its 110 years it had multiple locations within and near San José, so the exact placement of Chappe's observatory is unknown. But the latitude and longitude that Chappe's data produced places the crosshairs squarely on the small city itself—as close to spot-on as present evidence can determine.

After the lunar eclipse, however, Chappe's condition worsened. In desperation he tried to ride out of the mission on horseback but had to return. "He lay in a most deplorable condition, suffering the sharpest pains, and destitute of all assistance," wrote Pauly, who was fighting his own case of the fever. "The village . . . was by this time a mere desert. Three-fourths of the inhabitants were dead, and the rest had fled to seek a less infectious air. But the contagion had already spread far and wide."

Still, Chappe managed to record nine more days of solar observations to continue refining the accuracy of his timekeeper. In his logbook for

June 29, Chappe recorded the brief but revealing confession, "I observed, fatigued by lack of sleep and illness."[12] He stayed up another nine June and July nights, dutifully recording culmination measurements of the stars Arcturus and Kornephoros and immersion and emersion of Jupiter's moons.

The smell of death would have been inescapable, as well as the groans of sick and dying explorers and natives. Food remained available during this plentiful season, but picking and preparing, with no healthy servants to call on, was another matter. Chappe didn't have to venture far to get to his telescope, clock, and quadrant. But finding date palms near the mission whose fruit bunches hadn't already been completely picked apart was no small challenge for the rare healthy person in this pitiful realm of the dying. Climbing the trees—braving the knife-sharp basal leaves that guard the dates—was difficult enough. Now an unsteady hand, a weak body, and sometimes delirious mind made simple subsistence its own cruel daily punishment.

The jail fever chronicler described the last phase of the sickness. "Often either from neglect or an improper treatment in the beginning . . . [typhus] puts on a more fatal and alarming form. The pain of the head continues . . . the tongue as well as base of the teeth are covered with a thick black crust and the patient is unable to thrust it out of his mouth and loses the power of speech and of swallowing. . . . The pulse becomes weaker and quicker. . . . The patient is now altogether insensible. . . . He knows not the by-standers. . . . His muscles become flaccid. . . . He is affected with . . . convulsive startings and twitchings of the muscles. . . . His extremities become cold. A final quantity of blood sometimes distills from his nostrils. His face has a livid and cadaverous appearance, and death which soon follows these symptoms puts a period to his sufferings."[13]

On August 1, 1769, the forty-two-year-old Jean-Baptiste Chappe d'Auteroche drew his final breath. On his deathbed Chappe had said,

"I feel I have little time to live. But I have fulfilled my purpose, and I die happy."[14]

When he died, Chappe was surrounded by Pauly and Noël, whose own severe illnesses no doubt made them wonder how many days hence they might accompany their superior to the grave. "Doz and Medina did their best to pay last respects to Chappe for the priest was long since dead," Pauly said. "The Spanish, French and every one of the survivors then collected what little strength they had left and performed the most melancholy of all offices."

Chappe had asked Pauly to bury him in a Franciscan habit, a request that the morbidly bedraggled expedition honored. Crucially, Chappe had also charged Pauly with collecting his papers and ensuring their safe passage back to Paris. Fighting his own pain, delirium, and fever, Pauly gathered what journals and logbooks he could find and packed them in a casket, addressing it to the viceroy of New Spain. Pauly instructed one of the local chiefs to ensure the casket make it onto the ship that would be sailing for the mainland in September—should Pauly himself not survive the ensuing month and a half.

When word of Chappe's death reached Mexico City, Alzate—the polymath who had so delighted in the visiting Frenchman's presence five months before—was "greatly affected," Alzate wrote to the president of the French Royal Academy of Sciences. "New Spain has lost in him a man whose talents would have been of great service, to make known a thousand natural curiosities which here lie buried in oblivion."[15]

Chappe's colleagues at the Academy of Sciences in Paris could only concur. In the words of Chappe's eulogist, the academy's permanent secretary, the deceased "had an open and candid, unpretentious soul, and a noble, straightforward and honest heart; he was naturally lively, gregarious and amiable. He was known in the highest circles; the King himself deigned to converse with him and honored his death with expressions of regret. Never was there one more unselfish than he. He liked fame; he wished to earn its favors, not to steal them. . . . One could only have wished that the last proof he gave, so worthy of praise, had not been fatal to him."[16]

PARIS

1769–1770

Pauly and expedition artist Noël survived the cursed voyage, making them two of just nine (out of 28) who returned home. They made their way back to Paris via Mexico City—where the magnanimous viceroy had given them three expense-paid months to fully recover— and on to Vera Cruz and Cadiz. The voyage's scientific instruments, still property of the Royal Academy of Sciences, were returned to their rightful owners. Chappe's bereaved brother returned Ferdinand Berthoud's marine watch to the clockmaker. And Pauly put the contents of that casket full of papers and logbooks into the hands of France's Astronomer Royal, César-François Cassini de Thury.

Cassini, as he was known, edited Chappe's papers and published them three years later in *Voyage en Californie pour l'observation du passage de Vénus sur le disque du soleil*, a comprehensive volume commemorating the abbot's journey, reproducing his diaries and logbooks, and publishing the whole of Chappe's transit-related data. In the same 1772 volume, Cassini also wrote up his own history of the Venus transit and his summary of the data gathered worldwide on June 3, 1769, and its subsequent analysis.

The Venus transit expeditions of 1761, Cassini said, left a frustratingly vague answer to the ultimate question of the sun's distance. (Astronomers used a different number, the solar parallax, as a proxy for distance to the sun. See this book's Technical Appendix for more details.) "The result of the 1761 transit . . . enlarged the range of [possible parallaxes] from 8 1/2 arc seconds up to 10 1/2 arc seconds," Cassini wrote. "Thus the speculations of theory are found only too often belied by practice. One finds it very far indeed from the precision forecasted by Mr. Halley."[17]

Fortunately, Cassini noted, 1769 provided the world with a second chance. In fact, he calculated, the magnitude of the 1769 parallax—the difference in transit time at the North Pole compared to the equator—and the favorable locations available to observe it would not be duplicated again for a long time. The next three Venus transits—in 1874, 1882, and 2004—wouldn't offer nearly as propitious an opportunity to those distant future generations as did the most recent alignment of sun and planets. "It won't be until 2012 that the transit of Venus will be nearly as advantageous as it was in 1769," he said.

However, Cassini said, the many observers sent across the planet to witness the transit left behind an ocean of data. Some of it was good; some was not so good. Cassini argued, though, for concentrating on the most successful expeditions. "Three major voyages which, by their importance and usefulness, should be distinguished here: That of Father Hell to Vardø island, that of M. Chappe to California, and that of the English to the South Seas."

Because of his arrival at Vardø seven months before the transit, Cassini noted, Father Hell "had ample time to prepare and make a bountiful harvest of observations of different kinds that we hope to see in interesting detail in the considerable book this scholar promises us." Still smarting from the war of words between Hell and Lalande, Cassini laced his description of Hell's expedition with sarcasm—but conceded that Hell had at least provided all the detailed Venus transit data that the academy might require.

The data from Captain Cook's expedition, Cassini noted, constitutes "an observation whose success was so important to us and to serve as the standard of comparison of all the others. . . . [And] it cost the lives of those to whom we are indebted."

As for the voyage to California, "I will only say that M. Chappe's original destination was not California. It is infinitely desired that he went to the South Seas, the most favorable observing station. . . . The fruits that astronomy has derived from this observation have made it that much more precious and perpetuates forever the memory of the death of M. Chappe and of Don Salvador de Medina, one of the Spanish astronomers."[18]

Sifting through the data and competing parallax calculations, then, Cassini points out that the final number ultimately hinges on whether or not to reject Hell's transit observations.[19] Provisionally accepting Hell, Cassini says, enables five scientists to independently derive parallax figures that are remarkably close to one another: 8.7 arc seconds (Jérôme de Lalande), 8.7 arc seconds (Father Hell), 8.68 arc seconds (the Swiss mathematician Leonhard Euler), 8.76 arc seconds (French astronomy professor Jean Guillaume Wallot), and 8.88 arc seconds (astronomer Alexandre Guy Pingré).

These numbers are remarkably close to the correct value known today: 8.794 arc seconds. Technically, the ranking would be Wallot (99.6 percent of the exact result) followed by Pingré (99.0 percent), Lalande (98.9 percent), and Hell (98.9 percent). Euler—though undeniably the greatest genius of anyone remotely connected with the 1760s Venus transit voyages—nevertheless places last with a still laudable 98.7 percent. In all cases, the blinding view of hindsight suggests the story should end here.

However, because the imperfect results of the 1769 transit left open-ended questions, Cassini rejected every one of these calculations and instead settled on 8.5 arc seconds—adopting an arbitrarily rounded-off result that his colleague Lalande had arrived at. "Time, by which everything achieves its perfection, will clear up [the matters of] this article better than we can do," Cassini writes. "Until then, the average parallax of eight-and-a-half arc seconds, which had already been settled upon by the transit of 1761 [!], can be adopted, I think, without worrying too much about the truth."[20]

Cassini's arrogance in sweeping away such hard-won data—representing the greatest scientific cooperative endeavor in the history of mankind to date—is breathtaking. Fortunately, his pronouncement on the matter would not be the last.

Oxford

1771

Across the English Channel, Oxford University astronomy professor Thomas Hornsby—who had initially projected the best places in the world to observe the 1769 transit—also concentrated on data from Chappe, Hell, and Cook/Green in tabulating the solar parallax. Hornsby finessed the data too, recognizing that Cook made a more accurate observation of the transit's beginning (ingress) and that Cook and Green together satisfactorily observed the transit's end (egress). So Hornsby averaged Cook's and Green's egress result.

"The mean parallax will be found to be 8.78 (arc seconds), and if the semidiameter of the earth be supposed equal to 3,985 English miles, the mean distance of the earth to the sun will be 93,726,900 English miles," Hornsby stated. This is a number that Hornsby, unlike the pusillanimous Cassini, stuck with. And it was 99.8 percent accurate—exactly, as it happens, Halley's hoped-for result.[21]

One can find Hornsby's stunning paper today in the *Philosophical Transactions of the Royal Society* for 1771, sandwiched between accounts of "fossil alkaline salts" in Egypt and "basalt hills" in Germany. The many sciences the Royal Society explored—like its continental peers in Paris and St. Petersburg—were all slowly maturing, each with its own moments of clumsy beauty and scholarly derring-do.

But Hornsby's concluding observation, wrapped in the terse understatement that characterizes many of the greatest scientific papers, is almost like a dispatch from another age. At a time when eighty-four feet—the altitude of the first balloon flight, still a decade away—was a practically unimaginable distance from the ground, astronomy could nevertheless still slip the bonds of earth and measure hundreds of thousands of interplanetary miles with uncanny precision. Courtesy of characters such as Chappe, Hell, Sajnovics, Green, and Cook, inching a plumb line into outer space and discovering the layout of the entire near universe was now a simple commonplace.

"As the relative distances of the planets are well known, their absolute distances and consequently the dimensions of the Solar System will be as follows," Hornsby wrote. Here is his chart.

	Relative diſtance.	Abſolute diſtance.
Mercury,	387,10	36,281,700
Venus,	723,33	67,795,500
Earth,	1000,00	93,726,900
Mars,	1523,69	142,818;000
Jupiter,	5200,98	487,472,000
Saturn,	9540,07	894,162,000

All of Hornsby's calculated (absolute) planetary distances are accurate to between 99.2 percent and 99.6 percent of the correct values.

And somewhere, along the dimly demarcated road between Mars and Jupiter or Venus and Earth, one might imagine marker stones commemorating the lives sacrificed along this pathway to the stars. Here too passed frigates and carriages carrying men no greater than their unparalleled times—yet still measuring up to the greatest standards any space-voyaging civilization might hope to lay down.

EPILOGUE

When he calculated the dimensions of the solar system in 1771, Thomas Hornsby didn't know how close to spot-on he was. Nevertheless, Hornsby still claimed confidence with gusto. "The uncertainty as to the quantity of the sun's [distance]," Hornsby wrote, "deduced from the observations of the transit of Venus in 1761 . . . seems now to be entirely removed."[1]

Hornsby was technically right—at least about the transit data he found most trustworthy. Would that the story of the 1769 Venus transit simply ended here.

In fact, nearly 150 other observers also reported Venus transit data to London and Paris in 1769, 1770, and 1771. Some observers, such as Charles Green's brother-in-law William Wales who observed the transit in Hudson's Bay in present-day Canada, also performed superb observations in extremely challenging conditions. Others, however, were more gentleman dabbler than cutting-edge astronomer, adding their less-than-superior data to the swelling pool of Venus transit studies. Modern statistical analysis of the results would weight the amateur results less than those of the top expeditions and scientists. Statistical analysis was not so sophisticated at the time, however. (Hornsby gives a hint of such methods, albeit incompletely applied, when in his solar parallax calculation he averaged Cook and Green's transit times rather than considering them as independent observations.[2] Hornsby probably did so because Green's papers at the time of his death were in such disarray that Cook reported having to make "alterations" to Green's Venus transit data just to make it self-consistent.[3]) Journals and even newspapers of the time in Paris, London, Vienna, Philadelphia, St. Petersburg, and other cities across Europe and the New World published an estimated six hundred separate calculations based on different combinations of various transit data.[4]

Naturally, in the aftermath of the 1769 transit, the answer to the question, How far is the sun? depended on whom one asked. The French astronomer, Alexandre Guy Pingré, for instance, calculated a solar parallax of 8.88 arc seconds—99.0 percent of the correct value. Other prominent Swedish and Russian scientists calculated 8.43 and 8.63 arc seconds (95.8 percent and 98.1 percent accurate).[5]

Before long, delicate egos also entered into the calculations. Father Hell knew that his data from Vardø constituted the most reliable arctic Venus transit observations of any in the world. However, Hell had also withheld his results—indeed any formal announcement about the voyage itself—until he and Sajnovics could prepare a full report for the king of Denmark in February 1770. The *de facto* chairman of the world-wide 1769 Venus transit effort, Jérôme Lalande in Paris, on the other hand, may not have even *known* about Hell and Sajnovics's voyage before Hell's Feb. 1770 publication. In any event, Lalande had also grown all too accustomed to refining his calculations of the solar parallax without any access to the Vardø data. So when Hell's numbers finally arrived in Paris, Lalande viewed them with more suspicion than may have been warranted.[6]

It didn't help that both Lalande and Hell were prickly personalities whose sensibilities evidently bruised easily. Lalande wrote a lengthy review of Hell's data in the leading French scholarly journal of the time, *Journal des Sçavans.*[7] Lalande praised some of Hell's astronomical methods but harshly attacked others. Lalande found it laughable, for instance, that Hell didn't pay proper attention to the atmospheric density in the high arctic and—in Lalande's opinion—damaged his data as a result.

Nevertheless, once armed with Hell and Chappe's data, Lalande calculated a solar parallax very close to Hell's calculated parallax. Lalande found between 8.75 and 8.80 arc seconds. Hell found 8.70 arc seconds. All three solar parallaxes fall close to or within Edmund Halley's 99.2 percent projected accuracy of the Venus transit calculations. Again, posterity pleads the actors in the present melodrama to leave well enough alone.

They did not. Hell wrote a personal letter to Lalande imploring the Frenchman to exclude other arctic Venus transit measurements and use Hell's results instead. Naturally, Lalande took offense at Hell's overreach. But Lalande went too far in his response, turning around and publicly rejecting Hell's data entirely. In an April 1772 monograph, Lalande said a competing Swedish arctic Venus transit observation "has become the most important in all of Europe, serving as a comparison for all remote observations—with which it agrees completely." As if to poison the porridge for everyone, Lalande then arrived at a new solar parallax value of 8.5 arc seconds (reducing his accuracy of the solar parallax to 96.7 percent).

Lalande's ink-stained fit sent Hell into a fury, causing the snubbed Hungarian *pater* to launch a blistering attack at Lalande in a 116-page memoir of the entire Vardø transit observation. Rightly but for the wrong reasons, Hell urged scientists to ignore Lalande's fusillade and adopt a more reasonable solar parallax value of 8.70 arc seconds—based only on his data and the observations from Cook's voyage.[8] "Tahiti and Vardø will be the two columns upon which the true solar parallax of 8.70 [arc seconds] will rest firmly and be preserved," Hell wrote.[9]

Lalande and Hell's titanic clashes were only the most conspicuous example of the unintended, but inevitable, result of undertaking such a vast international science project with so many diverse collaborators. Different teams naturally had different opinions about every aspect of the international effort. And sorting it all out meant political as well as scientific gamesmanship. But no one had come up with a better political solution beyond the prevailing state political model of the day: absolute monarchy. Jérôme Lalande was, in effect, the king of Venus transits. And authority flowed down from him.

So Lalande's sniffy decision to declare the solar parallax 8.5 arc seconds carried more weight than it should have.

Some scientists dared venture their own estimates as low as 8.43 and as high as 8.80 arc seconds.[10] But it's no great surprise that as late as 1814, a popular account of the 1769 Venus transit voyages stated plainly that "some of the ablest astronomers in Europe . . . [determined] that the horizontal parallax of the sun is, at a mean, about 8 seconds and a half, and his distance from the earth, in round numbers, 95 millions of miles."

The American lecturer who made this statement, a popular Boston preacher named John Lathrop, sought to translate the heady transit data into an analogy comprehensible to his early-nineteenth-century audience. Lathrop said 95 million miles is a distance "so prodigious that a cannon ball going at the rate of 8 miles in a minute would be more than 22 years in traveling from our globe to the central and solar luminary of its orbit."[11]

The Venus transit voyages of 1761 and 1769 represent mankind's first international "big science" project—a familiar notion in an age of human genome projects, the Hubble space telescope, and the Large Hadron Collider. The transit projects' closest latter-day cousin, though, was another set of voyages that married advanced science and technology with extreme adventure: NASA's Apollo program to land men on the moon the 1960s and 1970s.

Some in the Apollo project considered themselves kindred spirits with explorer-scientists like Captain Cook. In fact the Apollo 15 mission— one with a strong scientific focus—so identified with Venus transit voyagers that the command module was named Endeavour and the astronauts actually brought with them into space a block of wood from the sternpost of the original HMS *Endeavour*.[12]

Equally significant, though, is the analogy between the mission and its consequences for technology and society. The worldwide race to build ever faster computers was already under way when President John F. Kennedy effectively launched the Apollo program on May 25, 1961. But Apollo would give the computer industry one of its most important early economic and technological boosts.

The integrated circuit, and with it industry titans like Intel, came of age when NASA provided vast pools of money and incentive to build the smallest, fastest, and lightest computer components imaginable.[13] It is no small coincidence that the same country that won the race to land men on the moon has in the decades since fostered a national industry— from UNIVAC and IBM to Apple, Microsoft, Google, and beyond— that remains the world's greatest designer, builder, and developer of computers and computer software.

So too with the Venus transit voyages of the 1760s and the military and geopolitical problem of finding navigation at sea. The technical battles pitting lunars against the nautical chronometer would have happened regardless of Venus's motions in the sky. But once the 1761 and 1769 transit voyages had overtaken the European imagination, summoning with it considerable royally sanctioned piles of money, projects like Nevil Maskelyne's world-changing *Nautical Almanac* became not only feasible but inevitable.

Although recent books and television coverage of the longitude race have put clockmaker John Harrison's genius rightfully in the spotlight, the pendulum of history now must swing back toward a slightly more complex reality. Harrison's chronometers will forever constitute an important piece of the longitude puzzle, especially in the nineteenth century, when the chronometers' greatest innovations could be cheaply and reliably mass manufactured. But from the 1760s through at least the 1820s, the lunar longitude methods that Maskelyne and his fellow Venus transit pioneers mastered remained a staple of British sea power and her vast and rapidly growing global empire—even assisting her former colonial possessions.

In 1822, for instance, a navigational handbook, *The American Practical Lunarian and Seaman's Guide*, quoted Pastor Lathorp's above enthusiastic description of the 1769 Venus transit—cannonball analogy and all. The *Guide* informed its readership that, more than two generations after Harrison had won the Longitude Prize, the pragmatic reality of navigation at sea still belonged to Maskelyne. "Every commander navigating a vessel to foreign ports should furnish himself with a good brass sextant, a *Nautical Almanac* and the requisite tables given in the epitomes of navigation," the *Guide* stated. "These are all that are required for the purpose of finding longitude."[14]

As reliable as Newton's laws, the law of unintended consequences holds true for the Venus transit voyages too.

Charles Mason and Jeremiah Dixon, first paired together to observe the 1761 Venus transit, had executed their job with such skill under such trying circumstances that the Royal Society employed the astronomers during the years 1763 through 1767 surveying disputed borders between the colonies of Maryland, Pennsylvania, Delaware, and Virginia. Neither Venus transit voyager would live to see the abbreviation of one of their names—Dixie—that would become synonymous with an entire region of America and, ultimately, emblematize the civil war of which Mason and Dixon's line marked the epicenter.

Captain Cook's Venus transit voyage had returned to such fanfare and acclaim that the British Admiralty soon commissioned a second

(1772–1775) and later third voyage (1776–1779) to follow swiftly on the heels of one another. Cook's three voyages, the last of which ended with his death in a battle on the beaches of Hawaii, changed the world like precious few nautical adventures since the days of Columbus. In addition to effectively opening the Pacific for broad and extended explorations, Cook had so mastered the battle against scurvy that it would never again pose an insurmountable problem for globe-spanning mariners.[15]

Even work done for Captain Cook that didn't succeed sometimes bred its own success. A chemist named Joseph Priestly noticed that Cook's scurvy-curing fresh vegetables and sauerkraut caused people to burp. Priestly hypothesized that perhaps liquids that caused the same reaction in a person's gut might also prevent scurvy. So he prepared apparatus to "impregnate water with fixed air" and shipped it out with Captain Cook on his second voyage. The "fixed air" (a.k.a. carbonated) beverage did not cure scurvy. But it was history's first prepared soft drink—a $370 billion industry today that makes an unlikely distant cousin to the Venus transit missions.

Jean-Baptiste Chappe d'Auteroche set in motion his own curious train of historical events. Though childless himself, Chappe's nephew Claude Chappe was so inspired by his uncle's historic voyages to Siberia and New Spain to observe the Venus transit that he devoted his life to science after reading *Voyage en Sibérie*.[16] Together with his three brothers, Claude Chappe invented the visual telegraph—a grid of mechanical semaphore relay stations that from the 1790s to the 1850s constituted Europe's first ever information network. At its peak the younger Chappe's telegraph system spanned more than three thousand miles across the continent with over 556 separate stations.[17]

The 1760s Venus transit voyages were also celebrated and immortalized in their day—recognized for a time as the scientific culmination of one of the Enlightenment's greatest decades. Charles Green's brother-in-law William Wales returned from his Hudson's Bay 1769 Venus transit voyage to find a commission as Captain Cook's new astronomer on the famous mariner's second voyage. In 1775, upon completing Cook's Pacific adventures, Wales then accepted an appointment as Master of Navigation Mathematics at Christ's Hospital School in London. Wales was a popular instructor who shared tremendous stories—and posed equally tremendous mathematical problems—concerning some of the most legendary odysseys of the day. One of Wales's star pupils was a sensitive boy who spent countless hours marveling over the "wonder and mystery of the universe"—and Wales's wonderful tales of Pacific and polar adventures.

The pupil was Samuel Taylor Coleridge, whose *Rime of the Ancient Mariner* distilled all he had learned from his Venus transit voyaging teacher into skeins of gossamer verse.[18]

Back in America, a theologian and Methodist minister set out to make a definitive edition of the Bible that would bring his age's greatest achievements in all fields of learning to the greatest story ever told. Adam Clarke, in his 1811 edition of the Bible, wrote a six-page footnote to Genesis 1:1, "In the beginning, God created the heavens and the Earth."

As Clarke noted, "The word heavens must therefore comprehend the whole solar system as it is very likely the whole of this was created in these six days." The minister then unspooled in his copious biblical footnote one and a half pages of astronomical tables (!) giving distances to sun and planets, as well as pertinent facts about the known satellites of all the planets. "The columns containing the mean distance of the planets from the sun," Clarke continued, "are such as result from the best observations of the two last transits of Venus, which gave the solar parallax to be equal to 8 [and] three-fifths of a degree."[19]

In 1830, American educator Hervey Wilbur wrote a popular book on astronomy. Even generations after the fact, the worldwide efforts to measure the 1761 and 1769 transits still commanded a tone of reverence. "The vast importance of correct observations of a transit of Venus is thus clearly seen," Wilbur wrote, "as it enables man to throw his measuring line through millions of miles in space and gauge the mighty dimensions of the sun shining in his strength."[20]

On June 5–6, 2012, the world will witness the final Venus transit of the twenty-first century. (Venus transits almost always come in pairs, separated by eight years. The century's other transit came on June 8, 2004.)

In North and Central America, Venus will begin to cross into the solar disk when the sun is low in the afternoon or early evening sky. The sun will have set already when the transit ends, some six-and-a-half hours later. Europe, eastern Africa, the Middle East, western Australia, and South Asia, will see the latter parts of the 2012 Venus transit as the sun rises on the morning of June 6. Japan, Indonesia, eastern China, Russia, and eastern Australia as well as Alaska, Hawaii, northwestern Canada, and islands in the Pacific Ocean will see the transit in its entirety.

These days, of course, the distance to the sun is well known—approx. 92,956,000 miles—and the value is more easily tracked and fine-tuned by radar measurements and other direct methods than by rare Venus transits. This is not to say, however, that Venus transits have ceased to be useful to astronomy.

The 2012 Venus transit may in fact help guide one field of cutting-edge science. One of the hottest research areas in astronomy is the discovery and study of "exoplanets," planets orbiting stars other than our own. At the time of this writing, 725 exoplanets have been discovered. New ones are being added to the tally practically every week. (The website http://exoplanet.eu keeps the latest tabs and statistics.)

A key motivating question in exoplanet research involves estimating the number of Earth-like planets in the Milky Way capable of harboring life as we know it. (We can only study exoplanets in our galaxy and nearby satellite star clusters. Other galaxies are too far away for present-generation telescopes to be able to perform such detailed observations of individual stars.) The first step toward an answer will be finding a bona fide "sister" Earth somewhere else in the galaxy. But once a planet closely resembling ours is found, astronomers and astrobiologists will want to know what the atmosphere around the sister Earth is like.

This is where present-day Venus transits come in handy. When Venus passes in front of the sun, the sun's light also passes through the tiny ring of Venus's atmosphere at the planet's outer edge. Astronomers routinely examine the spectrum of colors of light coming through their telescopes for hints of the distant object's composition and other characteristics. For instance, the atom hydrogen naturally gives off a red photon of light (with a wavelength of 656.3 nanometers) when its electron descends from its second excited orbital state to first excited orbital state. Tracking this "H-alpha" spectral line throughout stars and galaxies in every corner of the universe has long been a centerpiece of visual astronomy, crucially assisting the study of an object's motion, temperature, composition, age, and other properties.

Every atom and molecule has its own signature pattern of colors and spectral lines that it preferentially absorbs when light passes through it. So when Venus crosses the sun's face during a transit, the sun's light during that brief window contains tiny spectral signatures—absorption lines—emanating from Venus's atmosphere. The Venus transit, in other words, provides a rare but precious test case of an Earth-like planet with an atmosphere of known composition (thanks to Venus probes such as NASA's Venus Express and the former Soviet Union's Vega-1) as it passes in front of a well-studied star, our own.

In 2004, a team of nine French, Swiss, American, Spanish, and German astronomers observed the Venus transit, discovering not only the strong signature of carbon dioxide in the Venusian atmosphere but even the characteristic wind speeds at various altitudes above the planet's surface.[21] Not unlike the 1761 Venus transit, 2004 provided a proof of principle for the larger ideas that the twin transit eight years later might be able to test more fully.

In 2009, NASA launched the Kepler spacecraft, a dedicated exo-planet-finding, space-based telescope.[22] Kepler's *raison d'être* is to perform a kind of exoplanetary stakeout, continually keeping its telescope trained on a field of 145,000 stars in the constellations Cygnus, Lyra, and Draco. It primarily looks for exoplanets transiting their host stars. Astronomers then perform follow-up studies of the candidate exoplanetary systems that Kepler finds. Using observations such as the 2004 Venus transit data as reference points, they can then examine exoplanets' atmospheres—wherever such atmospheres might be found—in finer detail. Hints of life on the exoplanet could even turn up. The presence of O_2, molecular oxygen, in the Earth's atmosphere certainly gives indications of lifeforms on this planet.

In December 2011, nine American and French astronomers wrote a letter to *Astronomy & Astrophysics*, a prominent journal in the field, urging their colleagues to perform careful measurements of the century's other Venus transit. "The June 2012 transit will be a unique occasion," these scientists note, to study "a planet that was long believed to be the Earth's twin sister . . . [and] to discriminate between Earth-like and Venus-like atmospheres of exoplanets transiting their stars." [23]

As the two eighteenth-century Venus transits enabled science to grasp its place in the universe, so the pair of twenty-first century transits help turn up clues to some of the most fundamental questions in science today: Are we alone? If not, what is life like on other worlds scattered throughout the celestial beyond?

Important note for anyone attempting to view the 2012 Venus transit: *Do not view the sun without a proper solar filter on your eyeglasses, telescope, or binoculars. Viewing the sun directly without such protection can cause permanent eye damagae, even blindness.*

ACKNOWLEDGMENTS

This book represents a homecoming of sorts. I have devoted more than a decade to researching and writing about other subjects, not the least of which was Shakespeare and Elizabethan history. It's taken a while to circle back to my original field of study, physics and astronomy. So my thanks first go to the good professors—Joel Weisberg, Bill Titus, Bruce Thomas, Cindy Blaha, William Gerace, Art Swift, Mike Skrutskie, among many—who opened up a whole universe to this undergraduate and later graduate student. Their lessons continue to inspire.

On the other hand, science is only one small part of the 1760s Venus transit story. To historians Simon Schaffer, Per Pippin Aspaas, William Sheehan, John Westfall, and Stephen Wepster as well as Cliff Thornton and Wendy Wales of the Captain Cook Society, I owe a debt of gratitude for their kind assistance and feedback during the preparation of this manuscript. A central piece of the history—the foundation of chapters 7, 9, and 12—lay buried in a manuscript for which I only had the Hungarian translation. Translating the entire document affordably and promptly into English was an odyssey all its own. (This book, not unlike the voyagers whose story it tells, followed an unforgiving timetable laid down by a forthcoming Venus transit—in this case the June 2012 transit, by which time we needed to have our book on bookstore shelves.) I am grateful for the assistance that Zsuzsa Racz, Amanda Solymosi, Roy Wright Tekastiaks, Merilee Karr, Dave Goldman, Paul Olchvary, and Stephen desJardins each provided in the search. And Ilona Dénes, the translator whom I did hire, consistently provided superb work that was prompt, careful, and professional. My special thanks to her.

This book also drew heavily on the resources generously provided by a number of research libraries and librarians. Thanks to Adam Perkins at the Cambridge University Library and to the many helpful staff at the Smith College libraries (Neilsen, Hillyer, and Young Libraries), Forbes Library (Northampton, Mass.), the Amherst College Frost Library, the

University of Massachusetts Du Bois Library, Harvard University's Widener and Houghton Libraries and Yale's Sterling Library. In the earliest stages of publicizing and promoting this book, at the time of this writing, I have also benefited from the kind assistance of Meg Thatcher, Aliza Ansell, Gerit Quealy, Max Germer, Jennifer Margulis, Kris Bordessa, Lauren Ware, Brett Paesel, and Mona Gable.

This book simply would not have existed were it not for the patient, generous, and resourceful guidance of my literary agent, Jennifer Weltz, at the Jean V. Naggar Literary Agency. Her good grace and genial humor have been a wellspring of motivation and inspiration throughout this project. I am all the more grateful that her colleagues at JVNLA—Jessica Regel, Tara Hart, Laura Biagi, Elizabeth Evans, and Alice Tasman—have contributed their tremendous talents to this project as well. My editor, Robert Pigeon, has been a steadfast and true supporter of this project from his first exposure to the idea to its final production and beyond. I extend my thanks as well to the dedicated editorial, production, promotion, and marketing teams, at Da Capo and the Perseus Books Group—including Kevin Hannover, Timm Bryson, Sean Maher, Karstin Painter, and Lara Simpson Hrabota.

This book greatly benefited from the input of readers who graciously gave of their time to critique and provide crucial feedback on early drafts of the manuscript. For their helpful critiques and insights, I remain indebted to Burl Gilyard, Joe Eskola, Megan Eskola, Sabrina Feldman, Wendy Wales, Cliff Thornton, Danielle Dart, and Kirsten Jamsen.

My friends and family—especially my father and two children—have endured more deadlines and provided more spark and light in my life than I could have hoped to ask for. To them I am forever grateful. And, finally, I owe every good word to my wife, Penny, whose love and support throughout this often supremely challenging project has never flagged. Thank you.

TECHNICAL APPENDIX

This book requires no specialized mathematical or astronomical training to explore the human drama of the 1761 and 1769 Venus transit voyages. However, as a supplement to the discussion, the present appendix answers readers' curiosity about the specific methods the astronomers mentioned in this book used to calculate the solar distance. In so many words: *How, specifically, can one use the Venus transit to find the distance to the sun? And why wasn't there an easier way?*

The answer to the second question sets the stage for the first.

Every astronomical object in the sky appears as if projected onto a flat screen. Nothing immediately apparent about any star or planet might give the casual observer hints about its distance.

The primary reason ancient astronomers distinguished between stars and planets was that stars remained fixed in their positions with respect to one another, as observed night after night. Planets (from the Greek for "wanderer"), however, moved across the familiar stellar tapestry over the course of weeks and months.

From the time of the ancient Mayan, Chinese, Greek, and other early civilizations, astronomers, typically also serving as astrologers, devised elaborate theories to explain planets' wanderings—and what those wanderings might portend for kings or great events of the day.

After the Polish astronomer Nicolaus Copernicus (1473–1543) committed the ultimate heresy of replacing the earth with the sun as the center of the solar system, the German mathematician Johannes Kepler (1571–1630) made quantitative sense of Copernicus's framework. From a lifetime of studying detailed planetary observations by Danish astronomer Tycho Brahe (1546–1601) and others, Kepler ultimately derived three basic laws of planetary motion, which remain in use to this day.

Kepler's third law states that the square of the time a planet takes to complete one orbit of the sun is proportional to the cube of that planet's distance from the sun. In a simple equation:

$$P^2 = a^3 \qquad\qquad (1)$$

Where P is the planet's orbital period (the length of time, measured in earth years, the planet takes to complete one orbit) and a is the planet's average distance from the sun, measured in fractions of the earth-sun distance (astronomical units or AU). From well before Kepler's day, detailed charts of Venus's motions established that the planet completes one orbital period, one Venusian "year," every 0.615 Earth years.

Multiply 0.615 by itself and take the cube-root of the result to find Venus's distance from the sun as 0.72 AU.[1] But what, in practical distance units such as miles, is an AU?

Here is where astronomy remained stuck for more than a century.

Clever attempts to leverage precision measurements of Mars's and Mercury's orbits brought some astronomers in the late seventeenth and early eighteenth centuries close to answering the question.[2] But planets move slowly across the sky, and tracking their motions with respect to background stars was necessarily imprecise. Discovering the sun's distance required greater precision.

It's useful now to introduce an important term: the "solar parallax," not the solar distance, is actually the quantity astronomers sought. Solar parallax is an angular measurement, representing one-half of the angular size of the earth as seen from the sun. To use the analogy of a circle, the solar parallax is like the angular "radius" of the earth, as subtended from a distance of 1 AU. Fortunately, converting between solar parallax and solar distance is relatively easy. The distance to the sun is just the radius of the earth divided by the solar parallax. In real numbers, 92,956,000 miles = 3,963.2 miles ÷ 8.79414 arc seconds. (To do this on a calculator, an extra factor of 206,265 is needed to convert arc seconds to "radians," the natural unit of angular measurement.)

In 1716, soon to be Astronomer Royal Edmund Halley published an astronomical call to arms, revealing that for a brief window in June 1761 and again in June 1769, the planet Venus would be moving across a kind of interplanetary yardstick—the face of the sun.

Astronomers at different locations across earth could then time the duration of the Venus transit and compare answers (along with exact measurements of the observers' latitude) to triangulate the sun's distance.

Halley's method did *not* call for the observers to know their longitudes. And given how difficult longitude was to determine at the time, whether at sea or on land, Halley's method seemed to be a quick and easy route to the solar parallax, such a crucial number in science.

However, one of Halley's protégés, the French astronomer Joseph-Nicolas Delisle, examined the English method in closer detail and found it wanting. In 1761, for instance, the difference between the shortest and longest transit times would be just 13 minutes, making crucial the accuracy of each Venus transit duration measurement down to the second. It would also require the weather's cooperation for five or more hours. And the location of the longest transit time would be in the Indian ocean, notorious for its changeable weather conditions.

Delisle thus developed a supplementary method of discovering solar parallax from the Venus transit. Delisle's technique required an observer to mark the exact local time for just one of the four points of contact between Venus and the sun. (Those four points are the planet's outer and inner limbs touching the sun's edge on entry and exit—external and internal points of ingress and egress, respectively.) Delisle's method did only require observation of a single moment in the transit, thus making stations of multiple observers more likely to have at least one overlapping data point even in very uncooperative weather. Crucially, however, Delisle's method also required knowing both latitude and longitude of the observing station.

The story in the present book describes the compromise Halley-Delisle method that astronomers used in the 1761 and 1769 Venus transit voyages: measure latitude, longitude and as much of the Venus transit as possible. Astronomers after the fact would then use both Halley's and Delisle's methods to discover solar parallax, ideally enabling them to cross-check their results as well.[3]

For the purposes of the present appendix, we'll also consider two approaches to deriving solar parallax or distance. Neither is strictly the Halley or Delisle approach. Rather the first is much simpler—although far less accurate—than the other.

In both approaches, the basic idea is the same. Venus's silhouette follows different paths across the solar disk depending on where on the earth one observes the Venus transit. The greater the path length across the sun, the longer Venus will take to cross that path length. Observe Venus crossing the sun from two widely separated locations on earth, and the difference between the transit times will ultimately depend on three factors: the exact locations on earth where the astronomers observed the transit, the physical distance to Venus, and the distance to the sun.

Because of Kepler's third law, the proportional distance between Venus and the sun was well established. And if the observers independently determine their latitude and longitude, then the only remaining free variable is the distance to the sun.

The simpler approach, then, only requires some high school mathematics:[4]

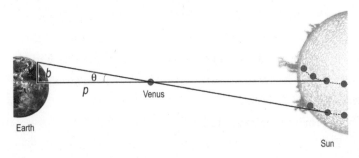

(relative sizes and distances not to scale)

Define the physical distance between two sets of Venus transit observers as b, and the distance between the earth and Venus as p. So by trigonometric definition—opposite over adjacent—the tangent of the apparent angular difference between the transit of Venus as seen by one observer compared to the other observer is:

$$\tan(\theta) = \frac{b}{p} \tag{2}$$

But the angle θ here is small, so $\tan(\theta) \approx \theta$. Therefore,

$$\theta = \frac{b}{p} \tag{3}$$

Multiply both sides of the equation by the ratio $\frac{a}{\theta}$, where a is the distance between the earth and the sun, and equation (3) becomes:

$$a = \left(\frac{b}{\theta}\right)\left(\frac{a}{p}\right) \tag{4}$$

And $\left(\frac{a}{p}\right)$ is already a known quantity, $\left(\frac{1}{1 - 0.72}\right)$, courtesy of Kepler's third law.

Ideally, one could take a time-lapse photograph of Venus as it traced a chord across the sun's face in, say, Vardø and compare that to the same time-lapse photograph of the Venus transit as seen by Chappe in Baja, Mexico, or by Cook and Green in Tahiti. The angular separation of the two Venus transits would be θ, and some smart cartography could yield b.

However, this approximation is too simplistic to do justice to the precision of the transit measurements and the needed precision of the calculations. For starters, it doesn't take into account the fact that b is measured over the curved surface of the earth. And, of course, there was no such thing as photography in 1769.

The approximation needs to be better.

Like many complex problems, there is no single correct way of arriving at an answer. Perhaps the most straightforward conceptually is the "educated guess" approach.

Namely, take the best result of the solar parallax from the 1761 Venus transits, and then blindly calculate the transit time one would expect to observe at any given latitude and longitude in the 1769 transit. Then factor *out* the estimated solar parallax and factor *in* the 1769 observations of transit times and known latitudes and longitudes.[5]

Mathematically, if solar parallax being calculated is π_{Sun} and the estimated solar parallax is π_{est}, then the ultimate expression would take the form:

$$\pi_{Sun} = \pi_{est} \frac{dt_O}{dt_C} \tag{5}$$

Where dt_O is the observed difference in Venus transit times, discussed below, and dt_C is the calculated difference in transit times.

Before getting into theoretical calculations, it's important to spell out precisely what data the three teams at the core of this book—Hell and Sajnovics at Vardø, Chappe at San José del Cabo, Cook and Green at Tahiti—brought home.

Observer	Place	Observed Latitude	Observed Longitude	Time of Inner Ingress, t_2 (local time)	Time of Inner Egress, t_3 (local time)	t_3-t_2
Hell	Vardø	70.367	31.02	$21^h33^m54^s$	$03^h27^m08^s$	$5^h53^m14^s$
Chappe	San José del Cabo	23.060	-109.68	$12^h17^m27^s$	$17^h54^m50^s$	$5^h37^m23^s$
Cook	Tahiti	-17.482	-149.48	$09^h44^m04^s$	$15^h14^m08^s$	$5^h30^m04^s$

Table A.1. Results of 1769 Venus transit expeditions as measured by teams led by Maximilian Hell, Jean-Baptiste Chappe d'Auteroche, and James Cook

So one of the two possible dt_Os here would be t_3-t_2 for Chappe subtracted from t_3-t_2 for Hell: $5^h53^m14^s$- $5^h37^m23^s = 15^m51^s$. The other would be t_3-t_2 for Cook subtracted from t_3-t_2 for Hell: $5^h53^m14^s$ - $5^h30^m04^s = 23^m10^s$.

What remains, then, is the calculation of dt_C, whose full derivation exceeds the scope of this appendix. Nevertheless, some essential equations and numerical results in these calculations can still yield a numerical answer.

It begins with so-called direction cosines of the three observers' locations on earth. Consider them x, y, and z components of unit vectors pointing to each observer's station (Vardø, San José del Cabo, Tahiti)—where this particular Cartesian coordinate system has its origin at the center of the earth and its x-y plane as the earth's equator and its x-z plane the Greenwich meridian.

Here, for example, are the direction cosines for Hell's observatory at Vardø:

$$\alpha_{Vardø} = \cos\phi_{Vardø} \cos\lambda_{Vardø} = (0.33599) \cdot (0.85699) = 0.28794$$
$$\beta_{Vardø} = \cos\phi_{Vardø} \sin\lambda_{Vardø} = (0.33599) \cdot (0.51534) = 0.17315 \quad (6)$$
$$\lambda_{Vardø} = \sin\phi_{Vardø} = 0.94186$$

Since the direction cosines describe unit-length arrows locating the various Venus transit observing stations, then similarly splitting up the dt_C calculation into its three Cartesian components might help to simplify this difficult problem. For instance, dt_C between Vardø and Tahiti would then take the form:

$$dt_c = A(\alpha_{Vardø} - \alpha_{Tahiti}) + B(\beta_{Vardø} - \beta_{Tahiti}) + C(\gamma_{Vardø} - \gamma_{Tahiti}) \quad (7)$$

Where the coefficients A, B, and C (whose units are in seconds) represent components of the differential transit time weighted to expected values of its x, y, and z contributions. The coefficients are not particular to any location on earth but rather to the geometry of the sun's position in the sky and Venus's typical path across the sun's face during the June 3, 1769, transit.

The (present-day) French astronomer François Mignard derived the first approximations of A, B, and C as follows—in this case, particular to each contact. (So for the following formulas, one would calculate separate coefficients for the 1769 transit's external ingress, internal ingress, internal egress, and external egress.)

$$A = \frac{\pi_{est}}{X\dot{X} + Y\dot{Y}} \left[\frac{1}{r_{Venus}} - \frac{1}{r_{Sun}} \right] (X \sin H_{Sun} - Y \cos H_{Sun} \sin \delta_{Sun} \quad (8)$$

$$B = \frac{\pi_{est}}{X\dot{X} + Y\dot{Y}} \left[\frac{1}{r_{Venus}} - \frac{1}{r_{Sun}} \right] (X \cos H_{Sun} - Y \sin H_{Sun} \sin \delta_{Sun} \quad (9)$$

$$C = \frac{\pi_{est}}{X\dot{X} + Y\dot{Y}} \left[\frac{1}{r_{Venus}} - \frac{1}{r_{Sun}} \right] (Y \cos \delta_{Sun}) \quad (10)$$

The variables X and Y (and their respective time derivatives) concern the position (and components of its speed) of Venus on the sun's disk, as seen through a telescope. The origin of the X-Y coordinate system is the sun's center, with X directed toward increasing right ascension[6] and Y toward celestial north. The values r_{Venus} and r_{Sun} represent the distances in AU from the earth to the respective bodies; thus 0.28 and 1. And H_{Sun} and δ_{Sun} are the right ascension and declination[7] of the sun as measured at Greenwich.

Despite the intimidating formulas above, there are just two key points to equations 8–10. First is the significance of any observer's additional Venus transit measurements—beyond timing, latitude, and longitude of the observatory. For example, Chappe's almost obsessive chronicling of every moment of Venus's transit—its angular speed across the sun's disk, its angular distance from the edge of the sun—would translate easily into $\dot{X}$, $\dot{Y}$, X and Y in the above formulas and thus help any solar parallax calculation enormously.

The second point is to demonstrate that A, B, and C are each directly proportional to the initial estimate of solar parallax from equation 5. So

π_{est} merely cancels out in this first approximation, leaving π_{Sun} independent of one's initial guess for the solar parallax.

The above represents only a first approximation of the coefficients A, B, and C, however. Mignard performed more detailed calculations of the three values and found—again, specific to the June 3, 1769, transit—$A = 476.5$ seconds, $B = 376.5$ seconds, and $C = 516.1$ seconds.[8]

At last we may be able to derive our own value of the solar parallax, π_{Sun}, from equations 5–7 and the data in Table A.1.

Sticking with the Vardø and Tahiti measurements, then, equations 6 and 7 yield:

$$dt_C = 476.5(1.1096) + 376.5(0.65753) + 516.1(1.24227) \text{ seconds} \quad (11)$$
$$= 1417.4 \text{ seconds} = 23_m37.4_s$$

Equation 5 along with the dt_O enumerated on p. 237 produces:

$$\pi_{Sun} = \pi_{est} \left(\frac{23^m 10^s}{23^m 37.4^s} \right) \quad (12)$$

Mignard used the modern-day value (8.794 arc seconds) for deriving his A, B, and C coefficients, which, as shown above, to first approximation is irrelevant to the final solar parallax calculation anyway. So, using just Hell's and Cook's 1769 Venus transit measurements and a reasonable facsimile of the information available to an eighteenth-century astronomer, we calculate a solar parallax value of:

$$\pi_{Sun} = 8.794 \left(\frac{1390}{1417.4} \right) = 8.62 \text{ arc seconds}$$

Note that this value is very close to related eighteenth-century calculations. In 1771, for instance, the Oxford University astronomy professor Thomas Hornsby used different equations to find that comparing Vardø-Tahiti data yields = 8.639 arc seconds.[9]

Finding the corresponding distance to the sun is just one more step, involving R_E, the radius of the earth:

$$D_{Sun} = \frac{R_E}{\pi_{Sun}} = \frac{3960 \; mi}{8.62"} = 94{,}750{,}000 \text{ miles.}[10] \qquad (13)$$

NOTES

Chapter 1: A Star in the Sun

1. Benjamin Martin, *Venus in the Sun* . . . (London: W. Owen, 1761), xi.

2. Council Minutes of the Royal Society, 4:254, cited in Harry Woolf, *The Transits of Venus* (Princeton: Princeton University Press, 1959), 85. The smartest mathematicians and philosophes of the day knew that vast improvements in the accuracy of measurement of the sun's distance might yield only marginal improvements in predicting the moon's and planets' positions in the skies months and years in advance (e.g., Tobias Mayer, discussed in S. A. Wepster, *Between Theory and Observations* [London: Springer, 2010]). Discovering the physical size scales of the solar system was the most pressing scientific problem in astronomy at the time, and astronomers—because of their ability to generate crucial nautical charts for mariners and admirals the world over—enjoyed no small degree of access to royal funding because of their solutions to the problems of longitude.

3. Johann Pezzl, "Sketch of Vienna," in H. C. Robbins Landon, ed., *Mozart and Vienna* (New York: Schirmer, 1991).

4. Don Michael Randel, "Joseph Haydn," in *The Harvard Biographical Dictionary of Music* (Cambridge: Harvard University Press, 1996), 367.

5. In the 1760s, outer space was considered the realm of giant, exalted objects like planets and stars. No one dared suggest that mundane things like rocks could be floating out there too. F. Brandstätter, "History of the Meteorite Collection of the Natural History Museum of Vienna," in *The History of Meteoritics and Key Meteorite Collections* (London: Geological Society, 2006), 123.

6. Van Swieten never convinced Chappe that electroshock therapy had any curative powers.

7. *Science, Technology, and Warfare: Proceedings of the Third Military History Symposium*, ed. Monte D. Wright and Lawrence J. Paszek (Honolulu: University Press of the Pacific, 2001), 78–79; Derek Edward Dawson

Beales, *Joseph II: In the Shadow of Maria Theresa, 1741–1780* (Cambridge: Cambridge University Press, 1987), 338; Bruce McConachy, "The Roots of Artillery Doctrine: Napoleonic Artillery Tactics Reconsidered," *Journal of Military History*, July 2001, 619–620.

8. Jean-Baptiste Chappe d'Auteroche, *A Journey into Siberia Made by Order of the King of France* (London: T. Jefferys, 1770), 25.

9. Chappe, *Journey*, 23.

10. Carol Jones Neuman, "The Historical Background," in *Drawings by Jean-Baptiste Le Prince for the Voyage en Sibérie* (Philadelphia: Rosenbach Museum and Library, 1986), 21–26.

11. Michel Mervaud, introduction to *Chappe d'Auteroche: Voyage en Sibérie fait par ordre du roi en 1761* (Oxford, U.K.: Voltaire Foundation, 2004), 6–15; Per Pippin Aspaas, private communication to author, January 3, 2012.

12. Chappe, *Journey*, 46.

13. Chappe, *Journey*, 47. Chappe says "the eldest [of the girls] was not above seventeen." Elsewhere he notes that girls in the Russian provinces were married at thirteen.

14. Chappe, *Journey*, 48.

15. Chappe, *Journey*, 49; Olga Yu. Elina, "Private Botanical Gardens in Russia: Between Noble Culture and Scientific Professionalization (1760s–1917)," in *The Global and the Local: The History of Science and the Cultural Integration of Europe: Proceedings of the Second ICESHS* (Cracow, 2006), www.2iceshs.cyfronet.pl.

16. Chappe, *Journey*, 52–53.

17. Chappe, *Journey*, 301.

18. Solar noon also coincides with the sun crossing an imaginary "meridian" line in the sky extending from directly north to directly south.

19. The present description is a facsimile of Chappe's actual latitude-calculating technique, chosen for its comparative simplicity. Technically, stellar charts give a star's "declination," the altitude of a star as measured from the "celestial equator," the projection of the earth's equator onto the sky. Calculating latitude via a star's altitude, then, involves first calculating the altitude of the celestial equator—which is the observed stellar altitude minus its declination, corrected for additional effects like atmospheric refraction. One's latitude, then, is 90 degrees minus the altitude of the celestial equator. This is the process Chappe actually followed. "Mr. L'Abbé Chappe d'Auteroche": *Memoire du passage de Venus sur le soleil* (St. Petersburg: Imperial Academy of the Sciences, 1762), http://tinyurl.com/262daq8. Translated by Mark Anderson.

20. Ibid. Astronomer Eric W. Elst discovered the present-day location of Chappe's observatory, at latitude 58 degrees, 11 arc minutes, 43 arc seconds; longitude 68 degrees, 15 arc minutes, 30 arc seconds. Chappe settled on 58 degrees, 12 arc minutes, and 22 arc seconds as his final answer (Ibid.). This represents a 39-arc-second difference between Chappe's calculation and modern determinations. Each arc second of latitude is 0.02 miles on the earth, which puts Chappe's error at 0.78 miles. www.holbach-foundation.org/astro/Details_obs.Chappe.htm.

21. Le Prince likely did not journey with Chappe to Tobolsk but instead traveled around Russia and Siberia separately. Upon returning to Paris, though, the artist did work in close collaboration with Chappe when preparing the drawings (on which the book's engravings would be based) for *Voyage en Sibérie*. Neuman, "Historical Background," 11.

22. Neuman, "Historical Background," 79–80.

Chapter 2: The Choicest Wonders

1. On the *Ramillies* wreck: http://sn.im/116pyp; www.submerged.co.uk/boltheadtobolttail.php.

2. "To the Author of London Magazine," *London Magazine, or, Gentleman's Monthly Intelligence* 30 (1761): 199, http://sn.im/11gqm0.

3. RS Misc. MSS 10/114. The East India Company was not a charitable organization. But every improvement in the art of navigation also improved its bottom line. The company's eagerness to ensure England had the best Venus transit measurements in the world bespeaks a recognition of some commercial potential in the science it was supporting.

4. The position of the planets and the moon was calculated using Starry Night Pro.

5. S. P. Rigaud, *Miscellaneous Works and Correspondence of the Rev. James Bradley* (Oxford, 1832), 388–390.

6. See http://sn.im/12xzim on the Byzantine role of white flags in British naval signaling. http://sn.im/12y07n; *Encyclopedia Britannica* (1797 ed.) on the white flag signaling "no hostile intention."

7. *A Universal Dictionary of the Marine*, "Rates," http://sn.im/126987.

8. *The New Bath Guide* (Bath: R. Cruttwell, 1789), 75–77, http://sn.im/198256.

9. Henry Francis Whitfield, *Plymouth and Devonport: In Times of War and Peace* (Plymouth: E. Chapple, 1900), 157.

10. RS Misc. MSS 10/128.

11. RS Misc. MSS 10/130.

12. RS Misc. MSS 10/129–131.

13. RS Misc. MSS 10/129.

14. James Pritchard, *Louis XV's Navy* (Montreal: McGill-Queen's University Press, 1987), 126.

15. RS Misc. MSS 10/134.

16. Table Bay descriptions: http://sn.im/19p306; http://sn.im/19p37h, 28ff.

17. Re Ryk (Rijk) Tulbagh, http://sn.im/19igqj; http://sn.im/19igs2.

18. Linnaeus, letter to Ryk Tulbagh, n.d., in *A Selection of the Correspondence of Linnaeus and Other Naturalists*, ed. James Edward Smith (London: Longman, Hurst, Rees, Orme & Brown, 1821), 2:570, http://sn.im/19qd40.

19. Abraham Bogaert (1702), quoted in Leonard Thompson, *A History of South Africa* (New Haven: Yale University Press, 1990), 39.

20. Kerry Ward, *Networks of Empire; Forced Migration in the Dutch East India Company* (Cambridge: Cambridge University Press, 2008).

21. Thompson, *History of South Africa*, 36–37; Hymen W.J. Picard, *Gentleman's Walk: The Romantic Story of Cape Town's Oldest Streets, Lanes, and Squares* (Cape Town: C. Struik, 1968), 255–128, http://sn.im/1ao3p0.

22. Theodore MacKenzie, in "Mason and Dixon at the Cape," *Monthly Notes of the Astronomical Society of South Africa* 10, no. 100 (1951), finds the observatory's location between present-day St. Johns and Hope streets near St. Mary's Cathedral. Using period maps (e.g. Picard, *Gentleman's Walk*, frontispiece and p. 16), the 1761 equivalent of this location can readily be found.

23. RS Misc. MSS 10/135.

24. RS Misc. MSS 10/143, 144.

25. Technically, the Venus transit observations ultimately yielded a number known as the solar parallax—one-half of the angular size of the earth, as seen from the sun. Because there are no stars visible in the daytime sky to measure the sun's position against, no one had ever been able to find out the solar parallax before Halley discovered the clever triangulation trick that enabled the Venus transit to yield the answer. The distance to the sun D can then be calculated from the equation $D = R/\pi$, where R is the earth's radius and π is the solar parallax. See the Technical Appendix in this book for more details.

26. Using Starry Night Pro set to Cape Town on June 5, 1761, one sees Antares rising nearly directly to the east at 8:40 PM. (Given the observatory's location, above, siting the mountains and landmarks above which it

would rise can readily be done in Google Earth.) Altair rises later, toward the northeast, at 11:40 PM.

27. Maskelyne, autobiographical notes, in Derek Howse, *Nevil Maskelyne: The Seaman's Astronomer* (Cambridge: Cambridge University Press, 1989), 216.

28. RS Misc. MSS 10/151.

29. "The Description and Use of Hadley's Octant, Commonly Called Hadley's Quadrant," in John Hamilton Moore, *The Practical Navigator* (London: B. Law & Son, 1791), http://sn.im/1dfkn3.

CHAPTER 3: FLYING BRIDGES

1. Guillaume-Thomas-François Raynal, *A Philosophical and Political History of the Settlements and Trade of the Europeans in the East and West Indies* (London: W. Strahan & T. Cadell, 1783), 8:190.

2. The archbishop, a notorious reactionary, had a few previous run-ins with Chappe. The archbishop hated the pope, whom he chastised for taking Communion sitting down. And he refused to believe that the earth orbited the sun. Chappe didn't sway the Russian prelate on the Copernican question. But as for the papal sacrilege, Chappe recalled, "I assured him the Pope was a cripple." Chappe, *Journey to Siberia*, 290.

3. William Tooke, *View of the Russian Empire During the Reign of Catherine the Second* (London: T.N. Longman & O. Rees, 1800), 2:45.

4. Russians still used the old Julian calendar. The new Gregorian date (in use across most of Europe) was January 19.

5. Chappe, *Siberia*, 338–340. James Boswell, *The Life of Samuel Johnson, LL.D.* (London: T. Cadell & W. Davies, 1816), 3:368–369. James Forfar, "The Czarina Elizabeth," *Gentleman's Magazine* 250 (1881): 606.

6. "The Life of Catherine II, Empress of Russia (review)," *European Magazine and London Review* 34 (1798): 390. History judges Peter III somewhat less harshly—partly as a victim of a propaganda campaign masterminded by his wife (and likely assassin), Catherine.

7. All quotes in Chappe's lecture are from "Mr. L'Abbé Chappe d'Auteroche," *Memoire du passage de Venus sur le soleil* (St. Petersburg: Imperial Academy of the Sciences), 1762. Translation by Mark Anderson.

8. Harry Woolf, *The Transits of Venus* (Princeton: Princeton University Press, 1959), 118–119, 145–147.

9. It was only in the nineteenth century, after the 1858 Treaty of Aigun, that the Russian empire would stretch to the Pacific.

10. Chappe, *Siberia*, 320–321.

11. Michael Reidy, Gary Kroll, and Erik Conway, *Exploration and Science: Social Impact and Interaction* (Santa Barbara, CA: ABC-CLIO, 2007), 19.

12. Richard Watkins, *Jérôme Lalande: Diary of a Trip to England, 1763* (Kingston, Tasmania: Richard Watkins, 2002), 161.

13. 3"/26" = 0.12 (dist. betw. eyes vs. length of arms), while 6356 x 2 km/ 150,000,000 km = 0.000085 (diam. of earth vs. AU, solar distance, or astronomical unit). Remember that the equivalent *parallax* distance for face would be nose-to-eye distance versus arm's length.

14. Woolf, *Transits*, 145.

15. Lalande's tendency toward snap judgments would one day cost him the discovery of the planet Neptune. Although he observed it in 1795, no one recognized it as a planet for another fifty-one years. Instead, Lalande reported that he observed just another "star," although his notebooks reveal he'd seen the position of the "star" shift over time. E. M. Standish, "Early Observations and Modern Ephemerides," *Highlights of Astronomy* 12 (2002): 327–328.

16. Nevil Maskelyne, "An Account of the Observations Made on the Transit of Venus, June 6, 1761, in the Island of St. Helena," *Philosophical Transactions of the Royal Society* 52 (December 1761): 199–200.

Chapter 4: The Mighty Dimensions

1. John Dobson, *Chronological Annals of the War* (Oxford: Clarendon, 1763), xiv, http://tinyurl.com/34qvxmy.

2. "From the top of the tower in Dunkirk, you can see the Thames." *Jérôme Lalande: Diary of a Trip to England, 1763*, trans. Richard Watkins (Kingston, Tasmania: Richard Watkins, 2002), 8, http://watkinsr.id.au /Lalande.pdf.

3. In "Fiction or Fact?" in *The Enigma of the Age: The Strange Story of Chevalier d'Eon* (London: Longmans, 1966) 11–25, Cynthia Cox sorts through claims and legends of d'Éon's possible *embassage en femme*. D'Éon ultimately developed such a contemporary reputation as a transgendered spy (to use a modern term) that in 1771 the London Stock Exchange sold financial instruments that effectively constituted bets as to whether d'Éon was a man or a woman. Jonathan Conlin, "The Strange Case of the Chevalier d'Eon," *History Today* 60, no. 4 (2010): 45–51.

4. Louis XV to d'Éon, June 3, 1763, in *Correspondence secrète inédité de Louis XV, sur la politique étrangère* (Paris, 1866), 1:293–294; Gary Kates, *Monsieur d'Eon Is a Woman* (New York: Basic, 1995), 93–94.

5. *Jérôme Lalande*, 12.

6. Owing to the ambassador's illness, D'Éon was serving a dual role as interim ambassador. *The Memoirs of Chevalier D'Éon*, trans. Antonia White (London: Anthony Blond, 1970), 123–124.

7. Contemporary description of the act in "The History of the Last Session of Parliament," *London Magazine* 32 (1763): 680–681, http://tinyurl .com/2brot2g.

8. David Alan Grier, *When Computers Were Human* (Princeton: Princeton University Press, 2005), 28.

9. *Jérôme Lalande*, 17.

10. *The European Magazine*, April 1783, 252, http://tinyurl.com /38kh4du.

11. *Jérôme Lalande*, 20.

12. "The King does not live in Kensington *at present*" (emphasis added). *Jérôme Lalande*, 20.

13. Charles Eyre Pascoe, *London of To-Day* (Boston: Roberts Bros., 1890), 324, http://tinyurl.com/2b6dv3t.

14. Nevil Maskelyne, *The British Mariner's Guide* . . . (London: Nevil Maskelyne, 1763), i–ii.

15. Maskelyne, *British Mariner's Guide*, v; (Jérôme) Lalande, *Connoissance des mouvemens célestes pour l'année 1762* (Paris: L'Impreimerie Royale, 1760), i, http://books.google.com/books?id=gJ0AAAAAMAAJ. Translation by Mark Anderson.

16. *Jérôme Lalande*, 13.

17. Maskelyne, *British Mariner's Guide*, iii.

18. Review of *The British Mariner's Guide* in *The Monthly Review* 28 (May 1763): 406, http://tinyurl.com/6ywvjsb. By contrast, see *Gentleman's Magazine* 43 (1773): 228–229, http://tinyurl.com/4w2o5n6, for a vitriolic critique of the *Mariner's Guide*.

19. Derek Howse, "The Lunar-Distance Method of Measuring Longitude," in *The Quest for Longitude*, ed. William J.H. Andrewes (Cambridge: Harvard University Press, 1996), 156.

20. Derek Howse, *Greenwich Time* (Oxford: Oxford University Press, 1980), 66–67.

21. *British Palladium* 12 (1764): 110; *Monthly Chronologer* 28 (September 1759): 505; *Annual Register* 10 (1761): 138, http://tinyurl.com /6hgg9r5; http://tinyurl.com/6yavmce; http://tinyurl.com/6jcumcr.

22. *Jérôme Lalande*, 29–30; Derek Howse and Anita McConnell, "Jeremiah Sisson," in *Oxford Dictionary of National Biography* (Oxford: Oxford University Press, 2004).

23. Isaac Newton, letter to Josiah Burchett, secretary of the Admiralty, August 26, 1725, in *Correspondence of Isaac Newton*, ed. H. W. Turnbull et al. (Cambridge: Cambridge University Press, 1959–1977), 7:330–332.

24. "Id vero an ipsi Daemone possible nescio." David S. Landes, *Revolution in Time* (Cambridge: Harvard University Press, 1983), 166–167.

25. On the "isochronal corrector" and "bimetallic strip," see William J.H. Andrewes, "Even Newton Could Be Wrong," in *Quest for Longitude*, 217–219.

26. *Jérôme Lalande*, 30.

27. Anthony Randall, "The Timekeeper That Won the Longitude Prize," in *Quest for Longitude*, 244–245.

28. Nevil Maskelyne to Edmund Maskelyne, December 29, 1763. National Maritime Museum PST/76/f.100–1, in Derek Howse, *Nevil Maskelyne* (Cambridge: Cambridge University Press, 1989), 49.

29. Ibid.

30. John Oldmixon, *The British Empire in America* (London: J. Brotherton, 1741), 161–162.

31. Nevil Maskelyne, "Astronomical Observations Made at the Island of Barbados . . ." *Philosophical Transactions of the Royal Society*, December 20, 1764, 389–392, http://tinyurl.com/47lzfmp.

32. Howse, *Maskelyne*, 50–51.

33. Randall, "Timekeeper," 247.

34. Wendy Wales, "Biography of Charles Green," *Cook's Log* 23, no. 4 (2000): 1775.

35. Sixty years later, the author and diplomat François René de Chateaubriand would look back to the 1763 treaty and wonder how "the government of my country would let perish her colonies that for us today would be an inexhaustible source of prosperity." Chateaubriand, *Voyage en Amérique* (Paris: Gabriel Roux, 1857), 219. Translation by Mark Anderson.

36. Jean-Baptiste Chappe d'Auteroche, remarks to the Académie Royale des Sciences, November 14, 1764, in Ferdinand Berthoud, *Traité des horloges marines . . .* (Paris: J.B.G. Musier, 1773), 539–541. Translation by Mark Anderson.

37. There has been some confusion on this point (e.g., Gould), but Berthoud says Chappe tested the "Montre Marine No. 3" in November 1764. Ferdinand Berthoud, *Traité des horloges marines, contenant la théorie, la construction, la main-d'oeuvre de ces machines, et la maniere de les éprouver* (Paris: J.B.G. Musier Fils, 1773), 539–545.

38. Although the Treaty of Paris may have ended the Seven Years' War, the king's admirals were still playing catch-up with the superior British fleet, preparing for the next inevitable go-round on the high seas. Jonathan

R. Dull, *The French Navy and the Seven Years' War* (Lincoln: University of Nebraska Press, 2007), 245.

39. Chappe, "Table de la marche de l'Horloge Marine" in Berthoud, *Traité*, 544.

40. Ibid., 542 fn.

41. Harrison's watch gained and lost seconds on a similar scale—but more predictably so. At 42 degrees Fahrenheit, the Harrison marine chronometer gained 3 seconds per 24 hours; at 52 degrees, 2 seconds; at 62, one second. These known inaccuracies could then be subtracted, yielding Harrison's groundbreaking results. Randall, "Timekeeper," 247 fn.

42. Chappe/Berthoud, *Traité*, 545.

CHAPTER 5: THE BOOK AND THE SHIP

1. S. A. Wepster, *Between Theory and Observations: Tobias Mayer's Explorations of Lunar Motion* (Springer: New York, 2010), 126.

2. Owen Gingerich and Barbara Welther, "Planetary, Lunar, and Solar Positions, A.D. 1650–1805," *Memoirs of the American Philosophical Society* 59S (1983): xxi, http://doiop.com/LunarsAccuracy.

3. Don't be deceived by the units. The earth rotates through 360 degrees every 24 hours—or 360/24 = 15 degrees every hour. So discovering that one lies, for instance, 3 "hours" from Greenwich means that Greenwich is 3 x 15 = 45 degrees of longitude away.

4. Mary Croarken, "Tabulating the Heavens: Computing the *Nautical Almanac* in 18th-Century England," *IEEE Annals of the History of Computing* 25, no. 3 (2003): 48–61.

5. E. G. R. Taylor, *The Haven-finding Art: A History of Navigation from Odysseus to Captain Cook* (London: Hollis & Carter, 1956), 263.

6. "St. Dunstan's in the West," in *London and Its Environs Described* (London: R. & J. Dodsley, 1761), 2:254–255; James Holbert Wilson, *Temple Bar: The City Golgotha* (London: David Bogue, 1853), 55–56.

7. *Philosophical Transactions of the Royal Society*, abridged ed. (London: C&R Baldwin, 1809), vol. 33 (1691), 448. The Royal Society began meeting at Crane Court in 1710. Halley was appointed Astronomer Royal in 1720.

8. Thomas Hornsby, "On the Transit of Venus in 1769 . . ." *Philosophical Transactions of the Royal Society* 55 (1765): 343–344.

9. John Black, *Travels Through Norway and Lapland During the Years 1806, 1807, and 1808* (London: Henry Colburn, 1813), 259.

10. Harry Woolf, *The Transits of Venus* (Princeton: Princeton University Press, 1959), 176. Bayley did obtain Venus transit data from nearby North Cape, but it was inferior to the Venus transit observations collected in

Vardø (at the invitation of the king of Denmark) by Father Maximilian Hell and Joannes Sajnovics, the subject of Chapters 7, 9, and 12 of the present book.

11. Royal Society Council Minute Book, 5:176–178, 181–200.

12. Hornsby calculated that the transit time observed in California would be some seventeen minutes different from the transit time observed in Lapland—providing data good enough to map out the whole solar system with the kind of 99.8 percent accuracy Edmund Halley had dreamed about.

13. Andrew Steinmetz, *The History of the Jesuits* (Philadelphia: Lea & Blanchard, 1848), 2:463.

14. RS: CMB, 5:183.

15. RS: CMB, 5:184–198.

16. RS: CMB, 5:282–285.

17. Brian Lavery, "Slade, Sir Thomas (1703/4–1771)," in *Oxford Dictionary of National Biography*, 2004; Brian Lavery, *The 74-gun Ship* Bellona (London: Conway Maritime Press, 1985), 7–9.

18. Alexander Dalrymple, "Memoirs of Alexander Dalrymple," *European Magazine and London Review* 42 (November 1802): 325.

19. Dalrymple, *Account* (London, 1767), ii–iv.

20. Dalrymple, an easy target, has nevertheless been slandered by history. In his defense, see Howard T. Fry, "Alexander Dalrymple and Captain Cook: The Creative Interplay of Two Careers," in *Captain James Cook and His Times*, ed. Robin Fisher and Hugh Johnston (Seattle: University of Washington Press, 1979), 44–47 fn. 17.

21. Patrick Brown, "The Civil and Natural History of Jamaica," *Monthly Review or Literary Journal* 15 (1756): 340.

22. Warrant Entry Book entry, cited in Arthur Kitson, *Captain James Cook* (New York: E.P. Dutton, 1907), 89.

23. Karl Heinz Marquardt, *Captain Cook's* Endeavour: *Anatomy of the Ship*, rev. ed. (London: Conway Maritime Press, 2001), 11–14.

24. Marquardt, *Captain Cook's* Endeavour, 14–18.

25. Lieutenant (later Captain) James Cook letter to John Walker, August 17, 1770, in Kitson, *Captain James Cook*, 219.

26. Philip Stephens, *Oxford Dictionary of National Biography* (Oxford: Oxford University Press, 2004), http://www.oxforddnb.com/index/26/101026391.

27. RSA: MSs (General) MS.633 in Andrew S. Cook, "James Cook and the Royal Society," in *Captain Cook: Explorations and Reassessments*, ed. Glyndwr Williams (Suffolk, UK: Boydell, 2004), 46 fn. 38.

28. RS: CMB, 5:299.

29. Andrew Kippis, *The Life of Captain James Cook* (Paris: J.J. Tourneisen, 1788), 1:209–212 fn.

30. It was a good thing, too. The *Aurora's* forthcoming passage to India would end in a gruesome shipwreck off the Cape of Good Hope. And in a crowning irony, the man who'd lobbied ultimately to die in Green's place, William Falconer, was the best-selling author of an epic seafaring poem. "The Shipwreck, by William Falconer," National Maritime Museum (UK), www.nmm.ac.uk/explore/collections/by-type/archive-and-library/item -of-the-month/previous/the-shipwreck,-by-william-falconer.

31. For example, *The Public Advertiser* (May 25–26, 1768), in Glyndwr Williams, "The *Endeavour* Voyage: A Coincidence of Motives," in *Science and Exploration in the Pacific*, ed. Margarette Lincoln (Suffolk, UK: Boydell & Brewer, 1998), 11.

32. *The London Magazine, or Gentleman's Monthly Intelligencer* 37 (June 1768): 328.

33. Wayne Orchiston, "James Cook's 1769 Transit of Venus Expedition to Tahiti," in *Transits of Venus: New Views of the Solar System and Galaxy*, ed. D. W. Kurtz (Cambridge: Cambridge University Press, 2005), 54; "An Account of Jesse Ramsden," *European Magazine and London Review* 15 (1789): 92.

34. Gowin Knight, in *Oxford Dictionary of National Biography*, http:// www.oxforddnb.com/index/15/101015719.

35. RS: CMB, 5:289–290.

36. James Bruce, *Travels to Discover the Source of the Nile* (Edinburgh: James Ballantyne, 1805), 7:361.

37. Patrick O'Brian, *Joseph Banks: A Life* (Boston: David R. Godine, 1993), 68–69.

38. George Robertson, *The Discovery of Tahiti: A Journal*, ed. Hugh Carrington (London: Hakluyt Society, 1948), 207–208.

39. Edward Smith, *The Life of Sir Joseph Banks* (London: John Lane, 1911), 15–16.

40. *The St. James's Chronicle* (June 11–14, 1768), in Williams, "*Endeavour* Voyage," 13.

41. *The Journals of Captain James Cook on His Voyages of Discovery*, ed. J. C. Beaglehole (Cambridge: Hakluyt Society/Cambridge University Press, 1955), 1:4. Entry for August 26, 1768.

42. *The Endeavour Journal of Joseph Banks, 1768–1771*, ed. J. C. Beaglehole (Sydney: Angus & Robertson, 1962), 1:158. Entry for September 10, 1768.

CHAPTER 6: VOYAGE EN CALIFORNIE

1. Chappe, *A Voyage to California: To Observe the Transit of Venus* (London: Edward & Charles Dilly, 1778) 13–14; James J. Fusco, "Abbé Chappe d'Auteroche: Eighteenth Century Modernizer," master's thesis, Columbia University, 1969, 13–14.

2. Andrew Dickson White, *The Warfare of Science* (New York: D. Appleton, 1876), 114–115; Johann Gottlieb Georgi, *Russia: Or, a Compleat Historical Account of All the Nations Which Compose That Empire* (London: J. Nichols, 1780), 3:355.

3. Benjamin Franklin, letter to Jean Chappe d'Auteroche, January 31, 1768 (15:33b), www.franklinpapers.org. Chappe's reply to Franklin, if any, has not been found. Chappe mentioned Franklin's correspondence in *A Journey into Siberia: Made by Order of the King of France* (London: T. Jefferys, 1770), 227.

4. Alexandre Guy Pingré, *Mémoire sur le choix et l'état des lieux où le passage de Vénus du 3 Juin 1769 pourra être observé avec le plus d'avantage* (Paris: P. Cavelier, 1767), 17. Translation by Mark Anderson.

5. Ibid., 78.

6. Académie Royale des Sciences, Proc. Verb. fol. 242 v. in Harry Woolf, *The Transits of Venus* (Princeton: Princeton University Press, 1959), 157 fn. 22.

7. Jean-Dominique de Cassini, "Histoire Abrégée de la Parallaxe du Soleil," in Chappe, *Voyage en Californie pour l'observation du passage de Vénus sur le disque du soleil* (Paris: Charles-Antoine Jombert, 1772), 155; translation by Mark Anderson; Woolf, *Transits*, 157.

8. Angus Armitage, "Chappe D'Auteroche: A Pathfinder for Astronomy," *Annals of Science* 10, no. 4 (1954): 288–291.

9. Jean-Dominique de Cassini, "Description of Cadiz," in Chappe, *Voyage to California*, 206.

10. Chappe, *Voyage to California*, 7–8.

11. Chappe, *Voyage to California*, 9, 15, citing Horace, *Horati Carmina* I.iii.9–11, trans. Samuel Maunder, in *The Treasury of Knowledge and Library of Reference* (New York: J.W. Bell, 1855), 2:100.

12. Water density, Chappe discovered, is not a useful proxy for longitude. It only changed substantially when the ship approached a freshwater source, such as a river that feeds into a bay.

13. Armitage, "Chappe D'Auteroche"; Chappe, *Voyage to California*, 11.

14. Vera Cruz's real traffic never approached the city gates. "In point of trade," one contemporary chronicler wrote, "[Vera Cruz] is one of the most considerable places not only in the New but perhaps in the whole world.

From this port it is that the great wealth of Mexico is poured out upon the old world." John Campbell, *An Account of the Spanish Settlements in America* (Edinburgh: A. Donaldson & J. Reid, 1762), 142. Vertiginous piles of Mexican gold and silver passed through Vera Cruz on their way to Spain—but only via a more navigable nearby island fortress, San Juan de Ulua, where Spain's plundered loot was stored and offloaded to awaiting galleons.

15. Chappe, *Voyage to California*, 15; "Vera-Cruz," in John Purdy, *The Columbian Navigator, or Sailing Dictionary* (London: R.H. Laurie, 1823), 128.

16. Campbell, *Account of the Spanish Settlements*, 141–142.

17. Ibid., 31–32.

18. Ibid., 36.

19. Fortunately the travelers found themselves on de Croix's good side. Church authorities leeward to the viceroy's favor had once demanded a meeting inside the cathedral. The former general arrived at the cathedral with an artillery detachment—informing his holy audience that the proceedings would need to be brief as his soldiers were under orders to reduce the building to rubble if de Croix didn't emerge from the church within ten minutes. Campbell, *Account of the Spanish Settlements*, 39–40; Henry Charles Lea, *The Inquisition in the Spanish Dependencies* (New York: Macmillan, 1908), 270 fn. 1.

20. Chappe, *Voyage to California*, 42–45; Lea, *Inquisition*, 272.

21. Chappe, *Voyage to California*, 77–105.

22. Ibid., 45–47.

23. Ibid., 47.

24. Steven Hales, "[On] the Late Earthquakes in London and Some Other Parts of England," *Philosophical Transactions of the Royal Society Abridged*, April 5, 1750, 10:539; Martin Uman, "Positive Lightning," in *The Lightning Discharge* (New York: Academic, 1987), 8–9, 188–204.

25. Chappe, *Voyage to California*, 49.

26. Ann Zwinger, *A Desert Country Near the Sea: A Natural History of the Cape Region of Baja California* (New York: Harper & Row, 1983), 111–119.

27. S. F. Cook, "The Extent and Significance of Disease Among the Indians of Baja California, 1697–1773," *Ibero-Americana* 12 (1937): 25–27.

CHAPTER 7: GREAT EXPEDITION

1. Joannes Sajnovics, travel diary entry for May 1, 1768, in *Sajnovics Naplója: 1768–1769–1770*, trans. Deák András, ed. Szíj Enikő (Budapest: ELTE Department of Finno-Ugric Linguistics, 1990). Translated

into English for the author by Ilona Dénes. This volume will hereafter be referred to as *Sajnovics's Travel Diary*.

2. Helge Kragh, *The Moon That Wasn't: The Saga of Venus' Spurious Satellite* (Berlin: Springer, 2008), 80–84.

3. Per Pippin Aspaas, "Maximilian Hell's Invitation to Norway," *Comm. in Asteroseismology* 149 (2008): 15.

4. Sajnovics, letter to unnamed correspondent, Vienna, April 16, 1768, *Sajnovics's Travel Diary*, 205.

5. Pinzger Ferenc, *Hell Miksa Emékezete . . . II. Rész* [To the Memory of Maximilian Hell, part II] (Budapest: Kiadja a Magyar Tudományos Akadémia, 1927): 31–32. Translated in Truls Lynne Hansen and Per Pippin Aspaas, "Maximilian Hell's Geomagnetic Observations in Norway 1769," *Tromsø Geophysical Observatory Reports* 2 (2005): 16.

6. *Sajnovics's Travel Diary*, May 3, 1768.

7. Ibid., May 4–6, 1768.

8. Ibid., May 30, 1768.

9. "The treaty by which Russia exchanged her claims on ducal Schleswig and Holstein for the counties of Oldenburg and Delmenhorst, which were intended to form an appanage for a junior branch of the Holstein family, was signed in 1768." C. F. Lascelles Wraxall, *Life and Times of Her Majesty Caroline Matilda* (London: Wm. H Allen, 1864), 130; W. F. Reddaway, "Don Sebastian de Llano and the Danish Revolution," *English Historical Review* 41, no. 161 (1926): 79.

10. *The Cambridge Modern History*, ed. A.W. Ward, G.W. Prothero, and Stanley Leathes (New York: Macmillan, 1909), 6:740–741.

11. *Sajnovics's Travel Diary*, May 31, 1768. Bernstorff had just turned fifty-six when Sajnovics and Hell met him.

12. On July 4, 1768, at Helsingor (Elsinore), the last Danish port of call before Hell and Sajnovics's party had to cross through Sweden en route to Norway, Sajnovics wrote that they had to leave Apropos behind "because it was forbidden to take dogs out of Denmark on account of a dangerous disease going around." However, the astronomers evidently spirited Apropos across the border anyway, as Hell's correspondence from Vardø includes mention of the dog's behavior. (Per Pippin Aspaas, personal communication with author, January 2, 2012.)

13. Maximilian Hell, letter to "Father Höller," April 6, 1769, www.kfki.hu /~tudtor/tallozo1/hell/hell2.html. Translated by Ilona Dénes.

14. Viggo Christiansen, *Christian den VII's Sindssygdom* (Copenhagen: Gyldendal, 1906), 62.

15. Wraxall, *Life*, 112–114; Karl Shaw, *Royal Babylon: The Alarming History of European Royalty* (New York: Broadway Books, 1999), 52–55.

16. Robert Gunning, quoted in W. F. Reddaway, "Struensee and the Fall of Bernstorff," *English Historical Review* 27, no. 106 (1912): 277.

17. *Sajnovics's Travel Diary*, July 22, 1768.

18. Sajnovics, letter from Nidrosa [Trondheim], August 1768, *Sajnovics's Travel Diary*.

19. Ibid., July 23, 1768.

20. Sajnovics, letter from Nidrosa, August 1768, *Sajnovics's Travel Diary*.

21. Ibid., July 25, 1768.

22. Ibid., July 28, 1768.

23. Ibid., July 26, 1768.

24. Sajnovics, letter, August 1768, *Sajnovics's Travel Diary*.

25. Sajnovics, letter from Haffnia [Copenhagen], June 21, 1768, *Sajnovics's Travel Diary*; Sajnovics, letter, August 1768, *Sajnovics's Travel Diary*.

26. In his diaries, Sajnovics variously spells Borchgrevink's name Purgreving, Purkreving, Purkgrevin, and so on.

27. *Sajnovics's Travel Diary*, August 8, 1768.

28. *Sajnovics's Travel Diary*, July 30, 1768.

29. Other officials played too. Sajnovics later wrote of the "concerts organized in our honor with a lot of mastery and artistry according to many. [Franz Joseph] Haydn, [Georg Christoph] Wagenseil, Gazmann and the other composers from Vienna are not entirely unknown here, and their creations are appreciated." Sajnovics, letter, August 1768, *Sajnovics's Travel Diary*.

30. *Sajnovics's Travel Diary*, August 22, 1768; Per Pippin Aspaas, personal communication to author, July 24, 2011.

31. Ibid.

32. "There were a few little ladies who also wanted to join us using all sorts of excuses," Sajnovics later recalled. "But Hell severely refused them." Sajnovics, letter from Vardø, November 14, 1768, *Sajnovics's Travel Diary*.

33. *Sajnovics's Travel Diary*, August 22, 1768.

34. Sajnovics, letter, November 14, 1768, *Sajnovics's Travel Diary*.

35. *Sajnovics's Travel Diary*, September 3, 18, 1768.

36. William Guthrie, *A New Geographical, Historical, and Commercial Grammar and Present State of the Several Kingdoms of the World* (London: Charles Dilly, 1794), 74.

37. Tobias George Smollett, *The Present State of All Nations: Containing a Geographical, Natural, Commercial, and Political History of All the Countries in the Known World* (London: R. Baldwin, 1768), 1:102–103.

38. *Sajnovics's Travel Diary*, September 25–26, 1768.

39. Sajnovics, letter, November 14, 1768, *Sajnovics's Travel Diary*.

40. *Sajnovics's Travel Diary*, October 7.

41. John Walker, *Elements of Geography and of Natural and Civil History* (Dublin: Thomas Morton Bates, 1797), 260; Jérôme (l'Abbé) Richard, *Histoire naturelle de l'air et des meteores* (Paris: Saillant & Nyon, 1770), 3:28.

42. Per Pippin Aspaas, personal communication to author, January 2, 2012.

43. *Sajnovics's Travel Diary*, October 13.

44. *Sajnovics's Travel Diary*, November 12.

45. NOAA Improved Sunrise/Sunset Calculator, www.srrb.noaa.gov /highlights/sunrise/sunrise.html.

46. Sajnovics, letter from Vardø, April 5, 1769, *Sajnovics's Travel Diary*.

CHAPTER 8: SOME UNFREQUENTED PART

1. Charles Green, manuscript, *Endeavour* journal for September 11, 1768, PRO Adm. 51/4545, f. 97.

2. *The* Endeavour *Journal of Joseph Banks, 1768–1771*, ed. J. C. Beaglehole (New South Wales: Angus & Robertson, 1962), 1:161.

3. Stephen Forwood, manuscript, *Endeavour* journal for September 14, 1768, PRO Adm. 51/4545, f. 231.

4. Charles Green, manuscript, September 14, 1768.

5. Endeavour *Journal of Joseph Banks*, 1:161.

6. Charles Green, manuscript, September 17, 1769, reports "complet[ing] our holds, having rece'd on board 3032 gallons of wine."

7. *The Journals of Captain James Cook on His Voyages of Discovery*, ed. J. C. Beaglehole (Cambridge: Cambridge University Press, 1955), 1:7–8.

8. William McBride, "'Normal' Medical Science and British Treatment of the Sea Scurvy, 1753–75," *J. Hist. Med. and Allied Sciences* 46 (1991): 167–169; R. Brookes, *The General Practice of Physic* (London: J. Newberry, 1765), 1:286.

9. Endeavour *Journal of Joseph Banks*, 1:163–164.

10. Anita McConnell, *Jesse Ramsden (1735–1800): London's Leading Scientific Instrument Maker* (Hampshire, UK: Ashgate, 2007), 159–160.

11. Joseph Priestley, *The History and Present State of Electricity, with Original Experiments* (London: J. Dodsley, 1767), 416.

12. Endeavour *Journal of Joseph Banks*, 1:160, 2:276–278.

13. E. G. Forbes, "The Foundation and Early Development of the Nautical Almanac," *J. Navigation* 18 (1965): 393–394; David Alan Grier, *When Computers Were Human* (Princeton: Princeton University Press, 2005), 30–33.

14. Maskelyne fired two of his "computers" when he discovered they were copying each other's results.

15. Forbes, "Foundation."

16. Derek Howse, "The Principal Scientific Instruments Taken on Captain Cook's Voyages of Exploration, 1768–80," *Mariner's Mirror* 65 (1979): 120–123; A. N. Stimson, "Some Board of Longitude Instruments in the Nineteenth Century," in *Nineteenth-Century Scientific Instruments and Their Makers* (Amsterdam: Rodpi, 1985), 109 fn. 20.

17. Green, manuscript, f. 105.

18. On his second and third voyages, Cook would also test some of Harrison's chronometers. *Astronomical Observations Made in the Voyages Which Were Undertaken by Order of His Present Majesty for Making Discoveries in the Southern Hemisphere*, ed. William Wales (London: C. Buckton, 1788), viii.

19. Mark Twain, *Following the Equator: A Journey Around the World* (Hartford, CT: American Publishing Company, 1898), 65.

20. Green, manuscript, ff. 107–111; Endeavour *Journal of Joseph Banks*, 1:170.

21. Fernando de Noronha's cornucopia of flora and fauna would have to wait to inspire Charles Darwin, who made it to the archipelago on his famous voyage on the HMS *Beagle*.

22. *Journals of Captain James Cook*, 1:16 fn. 2.

23. Endeavour *Journal of Joseph Banks*, 1:177.

24. Dauril Alden, *Royal Government in Colonial Brazil* (Berkeley: University of California Press, 1968), 110–111.

25. Cook, letter to the Royal Society, November 30, 1768, in Dan O'Sullivan, *In Search of Captain Cook: Exploring the Man Through His Own Words* (London: I. B. Tauris, 2008), 19.

26. Lieutenant Gore diary, cited in Nicholas Thomas, *Cook: The Extraordinary Voyages of Captain James Cook* (New York: Walker, 2003), 44.

27. *Journals of Captain James Cook*, 1:23.

28. Green, manuscript, f. 125.

29. Banks, letter to Lord Morton, in Patrick O'Brian, *Joseph Banks: A Life* (Boston: David R. Godine, 1993), 78.

30. Endeavour *Journal of Joseph Banks*, 1:190–191.

31. Richard Hough, *Captain James Cook* (New York: Norton, 1994), 68.

32. *Endeavour* journal by (prob.) James Magra in Thomas, *Cook*, 44.

33. Leslie Bethell, *Colonial Brazil* (Cambridge: Cambridge University Press, 1987), 257, 286.

34. Hough, *Captain James Cook*.

35. Alden, *Royal Government*, 109.

36. Thomas, *Cook*, 45; *Journals of Captain James Cook*, 1:31.

37. Banks had his own cabin but preferred to sleep on a hammock in the ship's great cabin.

38. Endeavour *Journal of Joseph Banks*, 1:212.

39. Wales, ed., *Astronomical Observations*, vii.

40. Endeavour *Journal of Joseph Banks*, 1:214.

41. Endeavour *Journal of Joseph Banks*, 1:217–218.

42. *Journals of Captain James Cook*, 1:45.

43. Endeavour *Journal of Joseph Banks*, 1:218–222.

Chapter 9: A Shining Band

1. Joannes Sajnovics, letter from Vardø, April 5, 1769, in *Sajnovics's Travel Diary*.

2. Sajnovics, letter from Vardø, April 5, 1769; Sajnovics, entry for December 23, 1768, *Sajnovics's Travel Diary*.

3. Barthold Georg Niebuhr, *The Life of Carsten Niebuhr: The Oriental Traveler*, trans. "Prof. Robinson" (Edinburgh: Thomas Clar, 1836), 39, http://tinyurl.com/hell-niebuhr.

4. The hard work had already been done. Hell discovered that the atmosphere at Vardø wasn't much thicker, so he could get away with using refraction tables already tabulated for the Royal Observatory in Paris. Truls Lynne Hansen and Per Pippin Aspaas, "Maximilian Hell's Geomagnetic Observations in Norway 1769," *Tromsø Geophysical Observatory Reports* 2 (2005): 15, http://geo.phys.uit.no/tgor/Hell-text.pdf.

5. Sajnovics, letter from Vardø, November 14, 1769, *Sajnovics's Travel Diary*.

6. Sajnovics, letter, April 5, 1769, *Sajnovics's Travel Diary*.

7. In 1770 Sajnovics wrote a book on the subject, *Demonstratio Idioma Ungarorum et Lapponum Idem Esse*, discussed below.

8. Maximilian Hell, letter to "Father Höller," April 6, 1769, Hungarian Electronic Library, www.kfki.hu/~tudtor/tallozo1/hell/hell2.html. Translated by Ilona Dénes.

9. Per Pippin Aspaas, "Maximilian Hell's Invitation to Norway," *Comm. in Asteroseismology* 149 (2008): 16.

10. Hell already knew gravity didn't pull precisely toward the earth's center, because of local variations in terrain and the earth's nonspherical imperfections. But he and Sajnovics spent nights during the dark months measuring stars' peak altitudes, which they could compare with star charts to better discover true vertical and thus their observatory's true horizon.

11. Vienna University Observatory, Manuscripte von Hell, *Observationes Astronomicae et caeterae Jn Jtinere litterario Vienna Wardoehusium usqve factae. 1768. A.M. Hell*, trans. Per Pippin Aspaas (personal communication to author, August 15, 2011); J. A. Bennett, *The Divided Circle: A History of Instruments for Astronomy, Navigation, and Surveying* (Oxford: Phaidon, 1987), 114–117.

12. Hell also had a feud with Niebuhr's mentor, Tobias Mayer, who championed the lunar method of determining longitude at sea. In England, Astronomer Royal Nevil Maskelyne based his *Nautical Almanac* on Mayer's work. But Hell remained skeptical of Mayer's methods. Niebuhr, *Life of Carsten Niebuhr*, 3:39; I. W. J. Hopkins, "The Maps of Carsten Niebuhr: 200 Years After," *Cartographic Journal* 4, no. 2 (1967): 115–118.

13. Sajnovics, letter from Vardø, June 6, 1769, *Sajnovics's Travel Diary.*

14. Smoked glass covered the opening to the telescopes, which enabled the observers to point their optics at such a luminous body as the sun and not completely burn their retinas.

15. For a detailed discussion of the Vardø transit, including assessment of reliability of each observer's data, see Simon Newcomb, "Wardhus," in *Astronomical Papers Prepared for the Use of the American Ephemeris and Nautical Almanac*, vol. 2, pt. 5 (Washington, D.C.: Bureau of Equipment and Recruiting, Navy Department, 1890), 301–305.

16. Maximilian Hell, *Observatio Transitus Veneris Ante Discum Solis Die 3 Junii Anno 1769* (Vienna: Joannis Thomae, 1770), 92. Translated by Bob Pigeon.

17. Ibid., 93.

18. "Te Deum," in *Encyclopedia Britannica*, 3rd ed. (Edinburgh: A. Bell & C. MacFarquhar, 1797), 18:332.

19. Sajnovics, letter, June 6, 1769, *Sajnovics's Travel Diary.*

CHAPTER 10: FORT VENUS

1. *The* Endeavor *Journal of Joseph Banks, 1768–1771*, ed. J. C. Beaglehole (New South Wales: Angus & Robertson, 1962), 1:222.

2. Charles Green, manuscript, PRO Adm. 51/4545, ff. 143–145.

3. *The Journals of Captain James Cook on his Voyages of Discovery*, ed. J. C. Beaglehole (Cambridge: Cambridge University Press, 1955), 1:55.

4. Endeavour *Journal of Joseph Banks*, 1:233.

5. John Clark, *Cook's* Endeavour *Journal: The Inside Story* (Canberra: National Library of Australia, 2008), 47–51.

6. Endeavour *Journal of Joseph Banks*, 1:242–243; Patrick O'Brian, *Joseph Banks: A Life* (Boston: David R. Godine, 1993), 86–87.

7. Today it's called Vahitahi.

8. Green, manuscript, f. 169.

9. *Journals of Captain James Cook*, 1:74. The role of sauerkraut in keeping scurvy at bay may be overplayed today, however, with Cook's fresh greens at each port of call playing at least an equal role in his mission's tremendous success at scurvy prophylaxis. Egon H. Kodicek and Frank G. Young, "Captain Cook and Scurvy," *Notes & Records of the Royal Society* 24, no. 1 (1969): 43–63.

10. *Journals of Captain James Cook*, 1:75–76.

11. David Howarth, *Tahiti: A Paradise Lost* (New York: Viking, 1983), 33–34.

12. Later Cook privately observed that the island's free and open sexual mores "can hardly be call'd a vice, since neither the state or individuals are the least injured by it." J. C. Beaglehole, *The Life of Captain Cook* (Stanford: Stanford University Press, 1974), 348.

13. Endeavour *Journal of Joseph Banks*, 1:252–258.

14. Cliff Thornton, personal communication with author, December 28, 2011.

15. Stolen metal goods often got incorporated into weapons, putting iron-armed Tahitian warriors at a distinct advantage against the usual blunt instruments rival factions on the island made from stones, shells, and shark teeth. O'Brian, *Joseph Banks*, 92.

16. Green, Manuscript, f. 278.

17. *Journals of Captain James Cook*, 1:87.

18. To the expedition's great fortune, Banks's assistant Herman Spöring was a former watchmaker and was able to perform whatever surgery the mauled quadrant demanded.

19. Endeavour *Journal of Joseph Banks*, 1:268–270; Richard Holmes, *The Age of Wonder: How the Romantic Generation Discovered the Beauty and Terror of Science* (New York: Pantheon, 2008), 5–7.

20. Derek Howse, "The Principal Scientific Instruments Taken on Captain Cook's Voyages of Exploration, 1768–80," *Mariner's Mirror* 65 (1979): 119–135.

21. Banks's assistant Solander had also set up his own three-foot reflecting telescope at Fort Venus for a redundant set of observations.

22. Wayne Orchiston, "James Cook's 1769 Transit of Venus Expedition to Tahiti," in *Transits of Venus: New Views of the Solar System and Galaxy*, ed. D. W. Kurtz (Cambridge: Cambridge University Press, 2004), 61.

23. *Journals of Captain James Cook*, 1:97–98.

24. Orchiston, "James Cook," 58.

25. Ibid., 57.

26. Endeavour *Journal of Joseph Banks*, 1:285.

Chapter 11: Behind the Sky

1. Salvador de Medina and Vicente de Doz, "Observations of the Transit of Venus," in *The 1769 Transit of Venus*, ed. Doyce B. Nunis, trans. Maynard J. Geiger (Los Angeles: Natural History Museum of Los Angeles, 1982), 121. In contrast, Chappe says (*Voyage to California*, 62) he spent the first night "determined not to go to [San José] till morning, [so] I laid me down by the water side."

2. Between 1730 and the mission's shuttering in 1840, there were actually multiple locations for Misíon Estero, two incarnations that were close to the water, one inland, one farther inland by up to eight kilometers. Circa 1769, the active mission was inland. There is no a priori reason to doubt the claim of "one mile from the beach" in Doz's account—although its exact whereabouts are unclear. Edward Vernon, *Las Misiones Antiguas: The Spanish Missions of Baja California, 1683–1855* (Santa Barbara, CA: Viejo, 2002).

3. The previous location on the shores of the gulf was abandoned in 1753 because of the terrible weather and swarms of mosquitoes that bred in the freshwater lagoon nearby.

4. On the Mexican mainland, by contrast, Jesuits had been running a corrupt network of missions, allegedly siphoning off the present-day equivalent of millions of dollars from their tills. Ann Zwinger, *A Desert Country Near the Sea: A Natural History of the Cape Region of Baja California* (New York: Harper & Row, 1983), 111–119.

5. Survey of Franciscan missions in Baja California, May 1773, in Charles Edward Chapman, *The Founding of Spanish California* (New York: Macmillan, 1913), 308–309.

6. S. F. Cook, "The Extent and Significance of Disease Among the Indians of Baja California, 1697–1773," *Ibero-Americana* 12 (1937): 25–27.

7. John Heysham, *An Account of the Jail Fever, or Typhus Carcerum, As It Appeared at Carlisle in the Year 1781* (London: T. Cadell, 1782), 8.

8. Joaquín Velázquez de Leon to Marqués de Croix, December 25, 1770, in *The 1769 Transit of Venus*, 133. Translated by Iris Wilson Engstrand.

9. Jean-Baptiste Chappe d'Auteroche, *A Voyage to California to Observe the Transit of Venus* (London: Edward & Charles Dilly, 1778), 63–65.

10. José de Gálvez to Fermín Francisco de Lasuén, November 23, 1768, in *The 1769 Transit of Venus*, 63.

11. The clockmaker Ferdinand Berthoud later wrote that he outfitted Chappe's expedition with one of his experimental spring-wound marine chronometers too. But finely crafted pendulum clocks still remained the most reliable timepieces on land. Ferdinand Berthoud, *Éclaircissemens sur l'invention, la théorie, la construction et les épreuves des nouvelles machines proposées en France pour la détermination des longitudes en mer* (Paris: J.B.G. Musier Fils, 1773), 25 fn. and especially 149.

12. Joaquín Velázquez to unknown colonial official, September 13, 1768, in Iris Wilson Engstrand, *Royal Officer in Baja California, 1768–1770: Joaquín Velázquez de León* (Los Angeles: Dawson's Book Shop, 1976), 45–46 fn.

13. "The Observations of Vincente de Doz," in *The 1769 Transit of Venus*, 122. Translated by Maynard J. Geiger.

14. *The 1769 Transit of Venus*, 123–124.

15. Heysham, *Account*, 9–10.

16. *The 1769 Transit of Venus*, 98–99.

17. See Chapter 9, note 15 about Hell's purported tenths of a second accuracy. Jean-Baptiste Chappe D'Auteroche, *Voyage en Californie pour l'observation du passage de Vénus sur le disque du soleil* (Paris: Charles-Antoine Jombert, 1772), 94. Translation by Mark Anderson.

18. "Observations of Vincente de Doz," 124.

19. Ibid., 125.

20. Ibid., 126.

21. *Voyage to California*, 65.

CHAPTER 12: SUBJECTS AND DISCOVERIES

1. Joannes Sajnovics, letter from Trondheim, September 2, 1769, *Sajnovics's Travel Diary*.

2. The Hungarian astronomer also recorded in his travel journal a similarly disappointing null result from Russian expeditions to observe the transit from the nearby Kola peninsula—one group coming up with nothing because of weather, the other because of what Sajnovics said was the sudden death of its lead observer. (However, reports of this Russian's death were indeed greatly exaggerated.) Entry for May 15 and June 14, *Sajnovics's Travel Diary*. Per Pippin Aspaas, personal communication with author, January 2, 2012.

3. Entry for October 19, 1769, *Sajnovics's Travel Diary.*

4. Per Pippin Aspaas, "Le Père Jésuite Maximilien Hell et ses relations avec Lalande," in *Jérôme Lalande (1732–1807): Une trajectoire scientifique* (Rennes, France: Presses Universitaires de Rennes, 2010). The author would like to thank Dr. Aspaas for sharing a prepublication version of the present article.

5. Entry for October 17, 1769, *Sajnovics's Travel Diary.*

6. Entry for December 8, 1769, *Sajnovics's Travel Diary.*

7. Letter to Count Bernstorff, October 1771, in W. F. Reddaway, "*Christian den VII's Sindssygdom* af Viggo Christiansen," *English Historical Review* 25, no. 97 (1910): 188–189.

8. Aspaas, "Lalande." De Luynes letter translated by Mark Anderson.

9. Sajnovics, letter from Copenhagen, February 10, 1770, *Sajnovics's Travel Diary.*

CHAPTER 13: SAIL TO THE SOUTHWARD

1. "Secret Instructions for Lieutenant James Cook Appointed to Command His Majesty's Bark the *Endeavour*," July 30, 1768 (NLA: MS 2), http://foundingdocs.gov.au/item-did-34.html.

2. *Endeavour Journal of Joseph Banks*, 1:312–313.

3. *Journals of Captain James Cook*, 1:155.

4. Endeavour *Journal of Joseph Banks*, 1:332–333.

5. *Journals of Captain James Cook*, 1:161.

6. A blind date with the Great Barrier Reef, for instance, would probably have ended in disaster but for *Endeavour*'s wonderfully unsexy flat keel. Karl Heinz Marquardt, *Captain Cook's* Endeavour: *Anatomy of the Ship*, rev. ed. (London: Conway Maritime Press, 2001), 11–14.

7. Brian W. Richardson, *Longitude and Empire: How Captain Cook's Voyages Changed the World* (Vancouver, B.C.: UBC Press, 2005), 73.

8. Patrick O'Brian, *Joseph Banks: A Life* (Boston: David R. Godine, 1993), 129.

9. *Journals of Captain James Cook*, 1:352.

10. Averil Lysaght, "Captain Cook's Kangaroo," *New Scientist*, March 14, 1957, 17–19.

11. James Cook, *Captain Cook's Voyages Round the World*, ed. M. B. Synge (London: Thomas Nelson, 1897), 148.

12. *Cook's Endeavour Journal: The Inside Story* (Canberra: National Library of Australia, 2008), 165–167.

13. *Journals of Captain James Cook*, 1:444.

14. Cliff Thornton, personal communication with author, December 28, 2011.

15. *Journals of Captain James Cook*, 1:448.

Chapter 14: Eclipse

1. John Heysham, *An Account of the Jail Fever, or Typhus Carcerum, As It Appeared at Carlisle in the Year 1781* (London: T. Cadell, 1782), 10.

2. Jean-Baptiste Chappe d'Auteroche, *A Voyage to California to Observe the Transit of Venus* (London: Edward & Charles Dilly, 1778), 67.

3. Chappe took thirteen such measurements during his stay at the mission, each of which helped fix the observatory's exact longitude—something the precision of his transit measurements now demanded.

4. Chappe, *Voyage to California*, 85.

5. The reason they're not exactly identical is that the earth is also moving approximately $1/365$ ($\approx 1°$) of its way around the sun on any given day, slightly shifting the field of view from star rise to star set. This offset, however, is predictable and can be subtracted out of any use of culminations to test a quadrant's accuracy—as the calculations show was done for Chappe's quadrant.

6. Chappe, *Voyage to California*, 75, 80–81, 85.

7. *The 1769 Transit of Venus*, ed. Doyce B. Nunis, trans. Maynard J. Geiger (Los Angeles: Natural History Museum of Los Angeles, 1982), 101; Chappe, *Voyage to California*, 67.

8. Chappe, *Voyage to California*, 68–69.

9. Chappe, *Voyage to California*, 89. Translation by Mark Anderson.

10. Ibid., 92.

11. Ibid., 88.

12. Nunis, *1769 Transit*, 93.

13. Heysham, *Jail Fever*, 11–12.

14. "Éloge de M. l'Abbé Chappe," in *Histoire de L'Académie Royale des Sciences* (Paris: 1772), 171. Translation by Mark Anderson.

15. Nunis, *1769 Transit*, 82.

16. Ibid., 87.

17. Chappe, *Voyage to California*, 150. Translation by Mark Anderson.

18. Ibid., 151–156.

19. Rumors still swirled around Hell's initial refusal to send the academy in Paris his Vardø transit data—and aroused anti-Jesuit tinged suspicions about the validity of Hell's findings.

20. Chappe, *Voyage to California*, 168. Translation by Mark Anderson.

21. The translation of parallax into physical distance from the sun also depends on an accurate value for the size of the earth, which was not precisely known in the eighteenth century. As a result, Hornsby's solar distance calculation loses a hair of precision—but still registers at an impressive 99.2 percent of the correct value.

Epilogue

1. *Philosophical Transactions of the Royal Society*, 61 (December 1771):574.

2. *Phil. Trans. R.S.* 61 (December 1771): 578; Harry Woolf, *The Transits of Venus* (Princeton: Princeton University Press, 1959), 190.

3. W. Orchiston, "James Cook's 1769 transit of Venus expedition to Tahiti" in *Transits of Venus: New Views of the Solar System and Galaxy* (Cambridge, Cambridge University Press, 2004), 58–61.

4. Woolf, *Transits of Venus*, 182–191.

5. Ibid., 190–191.

6. Per Pippin Aspaas, "Le Père Jésuite Maximilien Hell et ses relations avec Lalande," in *Jerôme Lalande (1732–1807) Une trajectoire scientifique* (Rennes, France: Presses Universitaires de Rennes, 2010).

7. *Journal des Sçavans*, September 1770, 619–622, in Aspaas, "Le Père."

8. Because Hell was a Jesuit, and therefore already suspect in the eyes of some, Hell and Sajnovics's entire mission was cast into eclipse. Until the late nineteenth century, many even suspected Hell of fabricating his expedition's Venus transit data. It was only with a careful study of Hell's manuscript archives in 1890 (Simon Newcomb, "Discussion of Transits of Venus, 1761–1769," in *Astronomical Papers Prepared for the Use of the American Ephemeris and Nautical Almanac* [Washington, D.C.: Dept. of Navy, 1890], 301–305) that Hell was vindicated and the allegations against his mission proved definitively wrong.

9. Hell, *Eph. Astr. anni 1773* (1772), in Aspaas, "Le Père."

10. Woolf, *Transits of Venus*, 192. As just one example, in 1779 A.I. Lexell used a solar parallax value of 8.63 arc seconds to calculate the mass ratio of the earth to the sun. C.A. Wilson, "Perturbations & Solar Tables," *Archive for History of Exact Sciences* 22 (1980) 195.

11. John Lathorp, "Lectures on Natural Philosophy (Lecture X)," *Polyanthos*, n.s., June 1814, 133–134.

12. Hamish Lindsay, *Tracking Apollo to the Moon* (London: Springer-Verlag, 2001), 316.

13. Sharon Gaudin, "NASA's Apollo Technology Has Changed History: Apollo Lunar Program Made a Staggering Contribution to High Tech Development," *Computerworld*, July 20, 2009.

14. Thomas Arnold, *The American Practical Lunarian and Seaman's Guide* (Philadelphia: Robert Desilver, 1822), 4:437.

15. Egon Kodicek and Frank Young, "Captain Cook and Scurvy," *Notes and Records of the Royal Society*, June 1969, 43–63.

16. "Reading this book greatly inspired him, and gave him a taste for the physical sciences. From this point on, all his studies, and even his pastimes, were focused on that subject." Abraham Chappe, quoted in *Journal de Paris*, February 1, 1805, in Gerard Holzman and Björn Pehrson, *The Early History of Data Networks* (New York: Wiley-IEEE Computer Society, 1994), 50.

17. Diana Hook and Jeremy Norman, *Origins of Cyberspace* (Norvato, CA: HistoryofScience.com, 2001), 179–180.

18. Rita Griffin-Short, "The Ancient Mariner and the Transit of Venus," *Endeavour*, December 2003, 175–179.

19. *The Holy Bible Containing the Old and New Testaments*, ed. Adam Clarke (New York: Ezra Sargeant, 1811).

20. Hervey Wilbur, *Elements of Astronomy, Descriptive and Physical* (New Haven, CT: Durrie & Peck, 1830), 83.

21. P. Hedelt, et. al. "Venus transit 2004: Illustrating the capability of exoplanet transmission spectroscopy." *Astronomy & Astrophysics*, vol. 533 (Sept. 2011) id. A136, http://arxiv.org/abs/1107.3700.

22. http://www.nasa.gov/mission_pages/kepler/main/index.html

23. David Ehrenreich et. al., "Transmission spectrum of Venus as a transiting exoplanet," *Astronomy & Astrophysics*, vol. 537 (Dec. 2011) id. L2, http://arxiv.org/abs/1112.0572.

Technical Appendix

1. Technically, planetary orbits describe an ellipse, a geometric figure that may be described nontechnically as a "squished circle." The distance *a* represents the so-called semi-major axis of the ellipse. However, for present purposes, the orbits of both earth and Venus are close to circular, so *a* closely approximates the average distance between planet and sun.

2. Albert van Helden, *Measuring the Universe: Cosmic Dimensions from Aristarchus to Halley* (Chicago: University of Chicago Press, 1985), 154–155.

3. William Sheehan and John Westfall, *The Transits of Venus*, (Amherst, N.Y.: Prometheus Books, 2004), 125–138. Raymond Haynes, *Explorers of the Southern Sky A History of Australian Astronomy* (Cambridge, U.K.: Cambridge University Press, 1996), 22–26.

4. Rezso Nagy and Attila Jozsef Kiss, "Observation of the Venus Transit," in *Jubilee Conference, 1879–2004* (Budapest: Budapest Tech Polytechnical Institution, 2004).

5. What follows is an adaptation of F. Mignard, "The Solar Parallax with the Transit of Venus," *Observatoire de la Côte d'Azur*, 2004.

6. "Right ascension" is an astronomical coordinate that represents the projection of terrestrial longitude on the sky.

7. "Declination" is the similar projection of terrestrial latitude on the sky. Ninety degrees declination is the north celestial pole, close to the location in the sky of Polaris, the North Star.

8. Ibid.

9. Thomas Hornsby, "The Quantity of the Sun's Parallax," *Phil. Trans. Roy. Soc.*, December 1771, 575.

10. The calculation requires that the parallax be measured in radians, not arc seconds. An extra factor of 206,265 arc seconds/radian was also applied to equation 13.

INDEX